Advanced Thin Film Materials for Photovoltaic Applications

Advanced Thin Film Materials for Photovoltaic Applications

Editor

I. M. Dharmadasa

MDPI • Basel • Beijing • Wuhan • Barcelona • Belgrade • Manchester • Tokyo • Cluj • Tianjin

Editor
I. M. Dharmadasa
Sheffield Hallam University
UK

Editorial Office
MDPI
St. Alban-Anlage 66
4052 Basel, Switzerland

This is a reprint of articles from the Special Issue published online in the open access journal *Coatings* (ISSN 2079-6412) (available at: https://www.mdpi.com/journal/coatings/special_issues/adv_thin_film_mater_Photovolt_appl).

For citation purposes, cite each article independently as indicated on the article page online and as indicated below:

LastName, A.A.; LastName, B.B.; LastName, C.C. Article Title. *Journal Name* **Year**, *Article Number*, Page Range.

ISBN 978-3-03943-040-6 (Hbk)
ISBN 978-3-03943-041-3 (PDF)

Contents

About the Editor

I. M. Dharmadasa, Ph.D., is the Head of the Electronic Materials & Sensors Group at the Materials and Engineering Research Institute at Sheffield Hallam University in the UK. He has four decades of experience in both industry (BP Research—Sunbury) and academia. His research focuses on the development of next generation, low-cost and high-efficiency solar cells using electroplated semiconductors. These solar cells based on CdS/CdTe currently show 15% efficient lab scale solar cells, and his efforts are focused on developing graded bandgap multi-layer solar cells. Prof. Dharmadasa has published over 250 articles, six patents, and two books on "Advances in Thin Film Solar Cells" and "Graded Bandgap Multi-Layer Solar Cells". In this process, he has successfully supervised 28 PhDs, 14 years of postdoctoral research and examined 32 doctoral candidates. He is also actively involved in the promotion of clean energy for the sustainable development and reduction of poverty. He has designed, piloted, monitored for several years and is now replicating the "Solar Village" project. He referees for over 12 journals and is one of the editors for two learned journals in his research field. He currently serves as an assessor/panel member for The European Commission and the Commonwealth Scholarship Commission.

 *coatings* 

Editorial

Special Issue: "Advanced Thin Film Materials for Photovoltaic Applications"

I. M. Dharmadasa

Materials and Engineering Research Institute, Sheffield Hallam University, Sheffield S1 1WB, UK; Dharme@shu.ac.uk

Received: 10 June 2020; Accepted: 11 June 2020; Published: 13 June 2020

Abstract: Photovoltaic (PV) technology is rapidly entering the energy market, providing clean energy for sustainable development in society, reducing air pollution. In order to accelerate the use of PV solar energy, both an improvement in conversion efficiency and reduction in manufacturing cost should be carried out continuously in the future. This can be achieved by the use of advanced thin film materials produced by low-cost growth techniques in novel device architectures. This effort intends to provide the latest research results on thin film photovoltaic solar energy materials in one place. This Special Issue presents the growth and characterisation of several PV solar energy materials using low-cost techniques to utilise in new device structures after optimisation. This will therefore provide specialists in the field with useful references and new insights into the subject. It is hoped that this common platform will serve as a stepping-stone for further development of this highly important field.

Keywords: thin films; perovskite; SnS/SnS$_2$; CdS/CdTe; CIGS; silicon; electroplating of semiconductors; photovoltaics

---

In the past, photovoltaic device development was mainly based on simple p-n homo- or hetero-junction type structures. However, these devices utilise only a fraction of the solar spectrum, and the rest is lost during the PV process. In order to harvest all photons from UV, Visible and IR regions, and add the contributions from "impurity PV effect" and "impact ionisation", graded bandgap multi-layer devices were designed [1]. These designs were experimentally tested using well known semiconductors (GaAs/AlGaAs), and their validity was proven by achieving V_{oc}~1.175 V and FF~0.86 [2]. After this validation, the new device architectures were fabricated using low-cost electroplated materials, and has achieved 15.3% efficiency to date. A monograph has been published [3] on this subject and the search for low-cost, advanced thin film materials is essential for the development of next-generation PV devices based on graded band-gap multi-layer solar cells.

This Special Issue consists of ten fully refereed scientific publications: seven open access articles [4–10] and three open access review articles [11–13]. The seven articles provide information on perovskite, SnS/SnS$_2$, CdTe, CIGS, silicon and transparent conducting oxide (SnO$_2$) materials used in solar cell development. One of these articles is a featured paper on electrodeposition of CdTe [7]. Out of the three review articles, one summarises the CdTe$_{(1-x)}$Se$_x$ thin films in solar cell applications [11]. The second review focuses on the encapsulation of organic and perovskite solar cells [12]. The third paper is a feature review of the electroplating of semiconductor materials for applications in large area electronic devices [13].

Among the research articles, Nishi et al. [4] present their latest work on CH$_3$NH$_3$PbI$_3$ perovskite material deposited under normal atmospheric conditions. These authors present devices with efficiencies ~14.3% and a stability up to four weeks, with the efficiency reducing only to 13.4%. This work shows the improvement in stability in the right direction. The next article by Gedi et al. [5] presents the results of eco-friendly SnS and SnS$_2$ thin films' growth and characterisation using chemical solution process. They report uniform and well-adhered layers with band gaps of 1.28 and 2.92 eV values, suitable for PV applications. Opyrchal et al. [6] report the photoluminescence study on the effect of Cu on the front side illumination of CdTe/CdS solar cells. The work focuses on the PL transitions close to the bandgap of CdTe. Ojo and Dharmadasa [7] present the results of electroplated CdTe material grown for use in CdS/CdTe solar cells. This article focuses on a case study of the temperature-dependent properties of electroplated CdTe thin films. Lorbada et al. [8], in their research article, provides a deep insight into the electronic properties of CIGS modules with monolithic interconnects. Chen et al. [9] present their results on enhancement of the potential-induced degradation resistance of crystalline silicon solar cells via anti-reflection coatings deposited by industrial PECVD method. The last research article by Ren et al. [10] presents the use of spin-coated SnO_2 thin films to cover cracks in the TiO_2 hole blocking layer used in perovskite solar cells. This process has improved the conversion efficiency of their solar cell structure.

The first review article by Lingg et al. [11] describes the properties of $CdTe_{(1-x)}Se_x$ thin films used in solar cell applications. First Solar Company has achieved ~22% efficient CdS/CdTe-based devices by incorporating Se in the CdTe layer. Hence, this comprehensive review is useful for researchers in this field to learn the properties of $CdTe_{(1-x)}Se_x$ alloy. The addition of Se in front of the solar cell creates a graded bandgap structure, enhancing the device performance. The second review paper by Uddin et al. [12] on the encapsulation of organic and perovskite solar cells is really important in order to improve the stability and lifetime of these types of solar cells. Although the highest thin film solar cell efficiencies of ~23% are reported for perovskite solar cells, their instability is a real concern at present. Hence, this encapsulation work is timely and useful for the researchers in this area. The last paper by Ojo and Dharmadasa [13] is a review paper on low-cost and high quality materials growth technique. This paper describes the electroplating of semiconductor materials for applications in large-area electronics such as PV solar panels and display devices. This will be an ideal paper for new researchers who intend to enter this area of research activities.

Finally, I would like to express my appreciation to all of the contributors to this Special Issue. They have positively responded to this call and their contributions are highly appreciated. Thanks are also due to the Coatings administration team for their efficient and excellent service, and for producing this professional publication.

Conflicts of Interest: The author declares no conflict of interest.

References

1. Dharmadasa, I.M. Third generation multi-layer tandem solar cells for achieving high conversion efficiencies. *Sol. Energy Mater. Sol. Cells* **2005**, *85*, 293–300. [CrossRef]
2. Dharmadasa, I.M.; Roberts, J.S.; Hill, G. Third generation multilayer graded bandgap solar cells for achieving high conversion efficiencies—II. *Sol. Energy Mater. Sol. Cells* **2005**, *88*, 413–422. [CrossRef]
3. Ojo, A.A.; Cranton, W.M.; Dharmadasa, I.M. *Next Generation Multi-Layer Graded Bandgap Solar Cells*; Springer: Berlin/Heidelberg, Germany, 2018.
4. Nishi, K.; Oku, T.; Kishimoto, T.; Ueoka, N.; Suzuki, A. Photovoltaic Characteristics of $CH_3NH_3PbI_3$ Perovskite Solar Cells Added with Ethylammonium Bromide and Formamidinium Iodide. *Coatings* **2020**, *10*, 410. [CrossRef]
5. Gedi, S.; Reddy, V.R.M.; Alhammadi, S.; Moon, D.; Seo, Y.; Kotte, T.R.R.; Park, C.; Kim, W.K. Effect of Thioacetamide Concentration on the Preparation of Single-Phase SnS and SnS_2 Thin Films for Optoelectronic Applications. *Coatings* **2019**, *9*, 632. [CrossRef]
6. Opyrchal, H.; Chen, D.; Cheng, Z.; Chin, K.K. PL Study on the Effect of Cu on the Front Side Luminescence of CdTe/CdS Solar Cells. *Coatings* **2019**, *9*, 435. [CrossRef]
7. Ojo, A.A.; Dharmadasa, I.M. Factors Affecting Electroplated Semiconductor Material Properties: The Case Study of Deposition Temperature on Cadmium Telluride. *Coatings* **2019**, *9*, 370. [CrossRef]

8. Lorbada, R.V.; Walter, T.; Marrón, D.F.; Lavrenko, T.; Mücke, D. A Deep Insight into the Electronic Properties of CIGS Modules with Monolithic Interconnects Based on 2D Simulations with TCAD. *Coatings* **2019**, *9*, 128. [CrossRef]

9. Chen, T.-C.; Kuo, T.-W.; Lin, Y.-L.; Ku, C.-H.; Yang, Z.-P.; Yu, I.-S. Enhancement for Potential-Induced Degradation Resistance of Crystalline Silicon Solar Cells via Anti-Reflection Coating by Industrial PECVD Methods. *Coatings* **2018**, *8*, 418. [CrossRef]

10. Ren, H.; Zou, X.; Cheng, J.; Ling, T.; Bai, X.; Chen, D. Facile Solution Spin-Coating SnO_2 Thin Film Covering Cracks of TiO_2 Hole Blocking Layer for Perovskite Solar Cells. *Coatings* **2018**, *8*, 314. [CrossRef]

11. Lingg, M.; Buecheler, S.; Tiwari, A.N. Review of $CdTe_{1-x}Se_x$ Thin Films in Solar Cell Applications. *Coatings* **2019**, *9*, 520. [CrossRef]

12. Uddin, A.; Upama, M.B.; Yi, H.; Duan, L. Encapsulation of Organic and Perovskite Solar Cells: A Review. *Coatings* **2019**, *9*, 65. [CrossRef]

13. Ojo, A.A.; Dharmadasa, I.M. Electroplating of Semiconductor Materials for Applications in Large Area Electronics: A Review. *Coatings* **2018**, *8*, 262. [CrossRef]

Article

Photovoltaic Characteristics of CH$_3$NH$_3$PbI$_3$ Perovskite Solar Cells Added with Ethylammonium Bromide and Formamidinium Iodide

Kousuke Nishi, Takeo Oku *, Taku Kishimoto, Naoki Ueoka and Atsushi Suzuki

Department of Materials Science, The University of Shiga Prefecture, 2500 Hassaka, Hikone, Shiga 522-8533, Japan; os21knishi@ec.usp.ac.jp (K.N.); oi21tkishimoto@ec.usp.ac.jp (T.K.); oh21nueoka@ec.usp.ac.jp (N.U.); suzuki@mat.usp.ac.jp (A.S.)
* Correspondence: oku@mat.usp.ac.jp; Tel.: +81-749-28-8368

Received: 1 April 2020; Accepted: 16 April 2020; Published: 20 April 2020

Abstract: Photovoltaic characteristics of solar cell devices in which ethylammonium (EA) and formamidinium (FA) were added to CH$_3$NH$_3$PbI$_3$ perovskite photoactive layers were investigated. The thin films for the devices were deposited by an ordinary spin-coating technique in ambient air, and the X-ray diffraction analysis revealed changes of the lattice constants, crystallite sizes and crystal orientations. By adding FA and EA, surface defects of the perovskite layer decreased, and the photoelectric parameters were improved. In addition, the highly (100) crystal orientations and device stabilities were improved by the EA and FA addition.

Keywords: perovskite solar cells; ethlammonium; formamidinium; microstructure

1. Introduction

Organic-inorganic perovskite solar cells provide photoelectric conversion in wide wavelength ranges and exhibit excellent photovoltaic properties [1–6]. Since the film of CH$_3$NH$_3$PbI$_3$ (MAPbI$_3$) can be formed by a spin-coating method, there is an advantage that the production process is easy and low cost. In spite of these merits, there is a serious problem that the stability is extremely low. In order to solve this problem, research and development of devices with higher power conversion efficiency and stability using formamidinium (FA) [7–13], guanidinium [14,15] or alkali metal [16–21] doped perovskites for the methylammonium (MA) site have been conducted.

There also exists research and development of devices with ethylammonium (EA) added to perovskites [22–26]. EA has a larger ionic radius (2.74 Å) than that of MA (2.17 Å), and the addition of EA can be expected to improve stability from the viewpoint of calculations [25,27] and tolerance factor [1]. In addition, there is a report that the thermal stability and crystallinity are higher than those of MA, and the addition of EA to the perovskites showed a surface coating with fewer defects and improves the stability of the device [23,28]. However, it should be noted that excessive addition of EA leads to phase separation, a decrease in crystallinity, and precipitation of PbI$_2$ as an impurity [29,30].

The purpose of this study is to examine the microstructures and photovoltaic characteristics of FA and EA co-added CH$_3$NH$_3$PbI$_3$ perovskite solar cells. The stability of a MAPbI$_3$ perovskite structure might be predicted by calculating the tolerance factor (t-factor) [31–35], which is given by $t = \frac{r_{MA}+r_I}{\sqrt{2}(r_{Pb}+r_I)}$, where r is an ionic radius [36]. When the t-factor is in the range of 0.81–1.1, perovskite structures could be formed [35]. If the t-factor is adjusted to 1.0, perovskite structures with cubic symmetry could be realized. The ionic radii of MA$^+$, FA$^+$, EA$^+$, Pb^{2+}, I$^-$, Br$^-$, and Cl$^-$ are 2.17, 2.53, 2.74, 1.19, 2.20, 1.96, and 1.81 Å, respectively [35,36]. By adding FA$^+$ and EA$^+$ with larger ionic radii than MA$^+$, t-factor gets closer to 1, and the stability is expected to be improved. In addition, EA addition is expected to promote the crystal growth and improve the stability of the device [23,28], and there are few reports on simultaneous addition of FA and EA to the perovskite layer. The effects of the simultaneous addition to the perovskite compounds were analyzed by microstructural and photovoltaic characterization.

2. Materials and Methods

A cross-section and deposition process of the present perovskite solar cells is summarized and shown in Figure 1. A fluorine-doped tin oxide (FTO, Nippon Sheet Glass Company, Ltd., Tokyo, Japan) substrate was dipped and washed in an ultrasonic washing machine using acetone twice and methanol once, and cleaned with flowing N_2. The 0.15 and 0.30 M precursor solutions of TiO_2 were prepared from 0.055 and 0.11 mL titanium disopropoxide bis (acetyl acetonate) (Sigma Aldrich, Tokyo, Japan) and 1-btanol (1.0 mL, Nacalai Tesque, Kyoto, Japan). The solutions were cast on the transparent FTO, and spin-coated at 3000 rpm for 30 s and heat-treated at 125 °C for 5 min [37–39]. The processes with 0.30 M precursor solutions were repeated twice. In order to form a dense electron transport TiO_2, the deposited samples were annealed at 550 °C for 30 min. The mesoporous TiO_2 layer was deposited with TiO_2 nanoparticles (P-25, Aerosil, Tokyo, Japan) and polyethylene glycol (Nacalai Tesque, Kyoto, Japan) in ultrapure water. The solution was blended with acethylacetone (20 µL, Fujifilm Wako Pure Chemical Corporation, Osaka, Japan) and triton-X-1001 (10 µL, Sigma Aldrich, Tokyo, Japan) for 30 min, and allowed to stand for 24 h to remove bubbles from the mixed solution. The prepared TiO_2 mixed solution was spin-coated at 5000 rpm for 30 s and annealed at 550 °C for 30 min, and a mesoporous TiO_2 layer was formed.

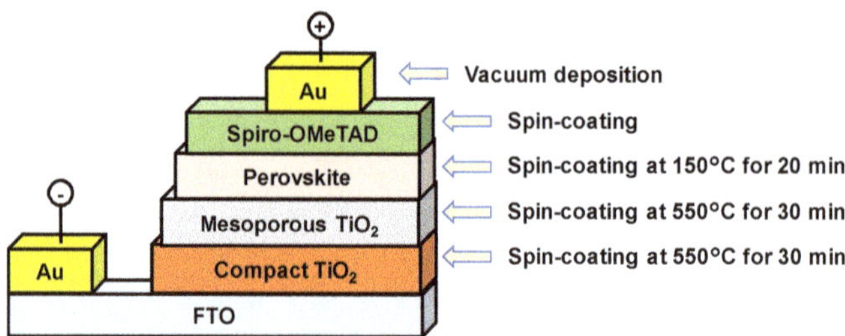

Figure 1. Cross-section of the cell and process conditions.

The perovskite precursor solutions were prepared as mixed solutions of methylamine hydroiodide CH_3NH_3I (MAI, 2.4 M, Tokyo Chemical Industry, Tokyo, Japan) and $PbCl_2$ (0.8 M, Sigma Aldrich, Tokyo, Japan) in N,N-dimethylformamide (DMF) (0.5 mL, Sigma Aldrich, Tokyo, Japan) at 60 °C for 24 h. This is used as a standard cell, and the amount of MAI was reduced by adding formamidine hydroiodide $CH(NH_2)_2I$ (FAI, Tokyo Chemical Industry, Tokyo, Japan), ethylamine hydrobromide $CH_3CH_2NH_3Br$ (EABr, Tokyo Chemical Industry, Tokyo, Japan), and ethylamine hydrochloride $CH_3CH_2NH_3Cl$ (EACl, Tokyo Chemical Industry, Tokyo, Japan). Detailed compositions of the perovskite compounds are listed in Table 1, together with the t-factors. The perovskite precursor solutions were spin-coated at 2000 rpm for 60 s and applied an air-blowing method during spin-coating [40,41]. The device was annealed at 150 °C for 20 min in the ambient air.

The hole-transport layer was deposited by spin-coating. A chlorobenzene solution (0.5 mL) of 2,2′,7,7′-tetrakis-(N,N-di-p-methoxyphenylamine)-9,9′-spirobifluorene (spiro-OMeTAD, Fujifilm Wako Pure Chemical, Corporation, Osaka, Japan, 36.1 mg) was prepared by mixing it for 12 h. An acetonitrile solution (0.5 mL) of lithium bis (trifluoromethylsulfonyl) imide (Li-TFSI, Tokyo Chemical Industry, Tokyo, Japan) was also prepared by mixing it for 12 h. A mixture solution of the spiro-OMeTAD solution with 4-tertbutylpridine (14.4 µL, Sigma Aldrich, Tokyo, Japan) and Li-TFSI solution (8.8 µL) was prepared by mixing it at 70 °C for 30 min. The spiro-OMeTAD layer was deposited by spin-coating at 4000 rpm for 30 s. After that, gold (Au) thin film electrodes were deposited as electrodes by vacuum evaporation. As investigated in the previous works [42–44], layer thicknesses of the compact TiO_2, mesoporous TiO_2 + perovskite, spiro-OMeTAD, and Au layers were roughly estimated to be 40, 600, 50, and 200 nm, respectively.

Table 1. Compositions and calculated *t*-factors of the present perovskite compounds.

Composition of Perovskite	EABr (%)	FAI (%)	*t*-Factor
$MAPbI_3$	0	0	0.912
$MA_{0.9}FA_{0.1}PbI_3$	0	10	0.919
$MA_{0.8}FA_{0.2}PbI_3$	0	20	0.927
$MA_{0.5}FA_{0.5}PbI_3$	0	50	0.949
$MA_{0.8}FA_{0.1}EA_{0.1}PbI_{2.7}Br_{0.3}$	10	10	0.933
$MA_{0.75}FA_{0.2}EA_{0.05}PbI_{2.85}Br_{0.15}$	5	20	0.933
$MA_{0.7}FA_{0.2}EA_{0.1}PbI_{2.7}Br_{0.3}$	10	20	0.940
$MA_{0.6}FA_{0.2}EA_{0.2}PbI_{2.4}Br_{0.6}$	20	20	0.954
$MA_{0.75}FA_{0.2}EA_{0.05}PbI_{2.85}Cl_{0.15}$	5	20	0.934
$MA_{0.7}FA_{0.2}EA_{0.1}PbI_{2.7}Cl_{0.3}$	10	20	0.941

The light-induced current density voltage (*J–V*) curves of the fabricated devices were obtained by using air mass 1.5 illuminator (San-ei Electric XES-301S, 100 mW·cm^{-2}) and a current-voltage apparatus (B2901A, Keysight, Santa Rosa, CA, USA). In addition, the external quantum efficiencies of the devices were obtained (QE-R, Enli Technology, Kaohsiung, Taiwan). Optical microscopy (Eclipse E600, Nikon, Tokyo, Japan) and X-ray diffraction (D2 PHASER, Bruker, Billerica, MA, USA) measurements were performed to analyze the surface morphologies and nanoscopic structures.

3. Results and Discussion

J–V curves collected in the light condition for the fabricated perovskite solar cells are displayed in Figure 2. Table 2 shows summarized parameters of the fabricated solar cells. A conversion efficiency (η) of the standard cell is 6.72%. The J_{SC}, V_{OC} and η were improved from 19.2 mA·cm^{-2}, 0.687 V and 6.72% to 21.5 mA·cm^{-2}, 0.922 V and 14.25% by addition of FA 20% at the MA site. When EA 10% and FA 10% were added simultaneously, the J_{SC}, V_{OC} and η increased 19.9 mA cm^{-2}, 0.946 V and 12.43%. Addition of EACl was also effective for the improvement of the device properties. Further addition of EA and FA would decrease the device performance.

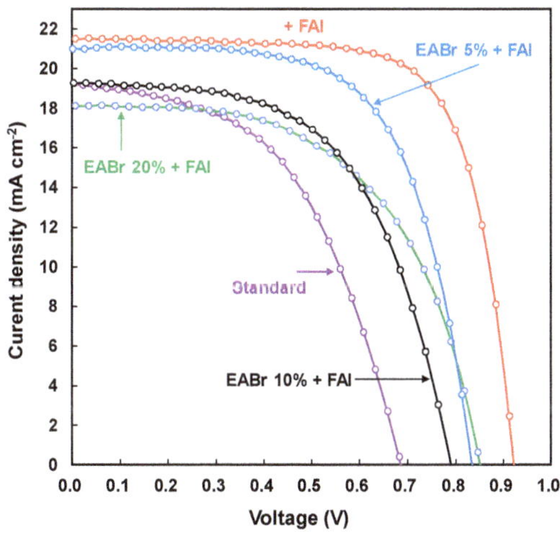

Figure 2. *J–V* characteristics collected in light condition for the fabricated solar cells.

Table 2. Measured parameters of the cells fabricated in this study. J_{SC}: short-circuit current density. V_{OC}: open-circuit voltage. FF: fill factor. R_S: series resistance. R_{Sh}: shunt resistance. η: conversion efficiency. η_{ave}: averaged efficiency of three cells.

Device	J_{SC} (mA·cm^{-2})	V_{OC} (V)	FF	R_S (Ω·cm^2)	R_{Sh} (Ω·cm^2)	η (%)	η_{ave} (%)
Standard	19.2	0.687	0.509	8.8	337	6.72	6.35
+FAI 10%	21.8	0.816	0.574	6.2	1663	10.24	8.04
+FAI 20%	21.5	0.922	0.719	3.4	4839	14.25	13.66
+FAI 50%	15.7	0.926	0.712	4.7	13,545	10.36	10.31
EABr 10% + FAI 10%	19.9	0.946	0.660	6.1	4667	12.43	12.23
EABr 5% + FAI 20%	21.0	0.834	0.648	5.6	4952	11.33	10.63
EABr 10% + FAI 20%	19.3	0.789	0.572	5.7	1015	8.47	8.70
EABr 20% + FAI 20%	18.1	0.851	0.562	4.8	2340	8.68	8.27
EACl 5% + FAI 20%	20.4	0.879	0.618	6.4	1879	11.06	10.63
EACl 10% + FAI 20%	20.2	0.933	0.647	5.2	66,637	12.21	11.64

Figure 3 is the *J–V* curves of the fabricated photovoltaic cells after 4 weeks in ambient air, and the estimated parameters are shown in Table 3. The conversion efficiency of the standard cell was lowered to 5.69%. Co-addition of small amount of EA and FA to MAPbI$_3$ provided higher stability compared with the standard cells, as shown in Figure 4.

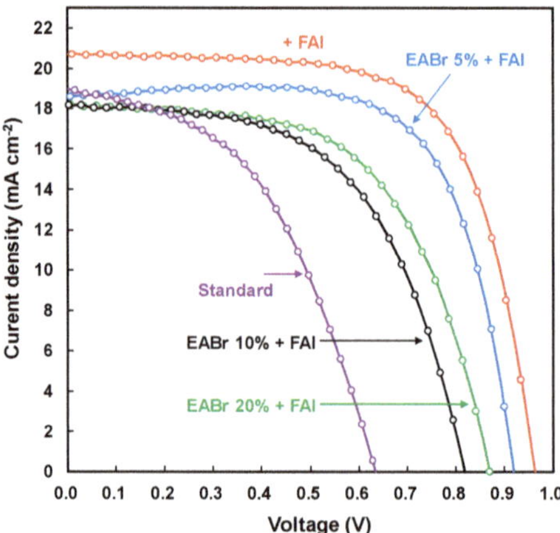

Figure 3. *J–V* characteristics collected in light condition for the fabricated solar cells after 4 weeks in ambient air without encapsulation.

Table 3. Measured photovoltaic parameters of the fabricated cells after 4 weeks.

Device	J_{SC} (mA·cm^{-2})	V_{OC} (V)	FF	R_S (Ω·cm^2)	R_{Sh} (Ω·cm^2)	η (%)	η_{ave} (%)
Standard	19.0	0.633	0.474	8.9	212	5.69	5.25
+FAI 10%	17.3	0.925	0.615	8.7	5123	9.85	9.30
+FAI 20%	20.7	0.961	0.675	4.6	2455	13.43	13.30
+FAI 50%	14.8	0.964	0.684	6.0	75,968	9.74	8.99
EABr 10% + FAI 10%	17.3	0.925	0.615	8.7	5123	9.85	9.30
EABr 5% + FAI 20%	18.6	0.919	0.699	5.2	19,971	11.93	11.41
EABr 10% + FAI 20%	18.2	0.819	0.564	7.6	1129	8.39	6.86
EABr 20% + FAI 20%	18.2	0.870	0.585	6.6	946	9.26	8.77
EACl 5% + FAI 20%	17.3	0.900	0.682	5.1	4097	10.62	9.98
EACl 10% + FAI 20%	17.0	0.932	0.664	5.8	5407	10.54	9.49

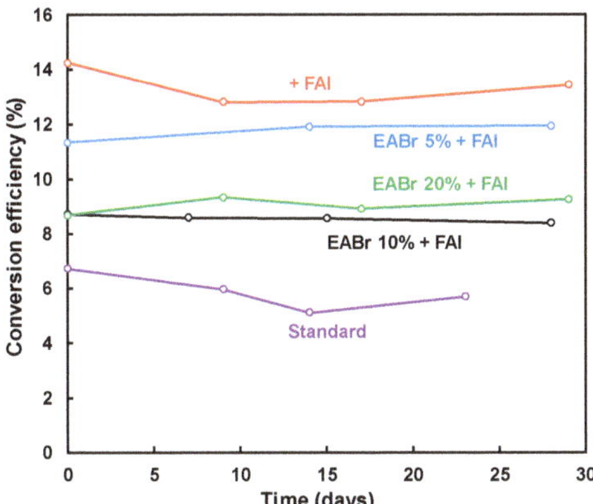

Figure 4. Stability measurements of the fabricated perovskite devices.

Optical microscopy images of the perovskites through spiro-OMeTAD are shown Figure 5. By adding EA and FA, surface defects of the perovskite layer decreased. Obtaining a perovskite layer with few defects enables efficient charge separation and charge extraction, which is thought to have led to improved device performance. In addition, defects in the perovskite layer are a cause of charge recombination, and it is considered that suppression of the defect has led to improvement in stability.

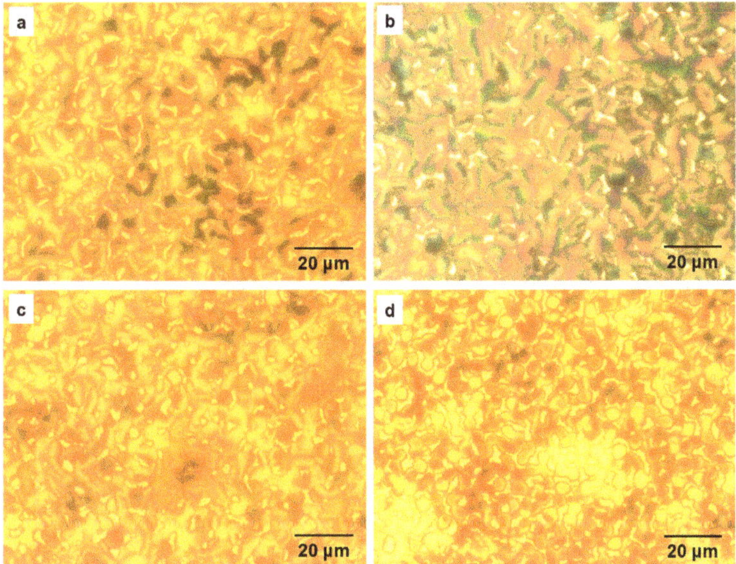

Figure 5. Optical microscopy images of cells with the compositions of (**a**) FAI 20%, (**b**) EABr 5% + FAI 20%, (**c**) EABr 10% + FAI 20%, and (**d**) EABr 20% + FAI 20%.

External quantum efficiency (EQE) spectra of the fabricated photovoltaic cells are shown in Figure 6. The band gap energies (E_g) were estimated from EQE spectra around 800 nm by linear fitting using band gap calculator software (Enli Technology, QE-R), and the measured band gap energies of the perovskite compounds increased from 1.54 to 1.57 eV by adding EA. The E_g value of the 20% EABr-added perovskite crystals was wider than that of the 20%FAI-added perovskite. The EQE values of the EABr-added device was lower between 350 and 750 nm than that of the FAI-added device, which led to a decrease of the J_{SC} values.

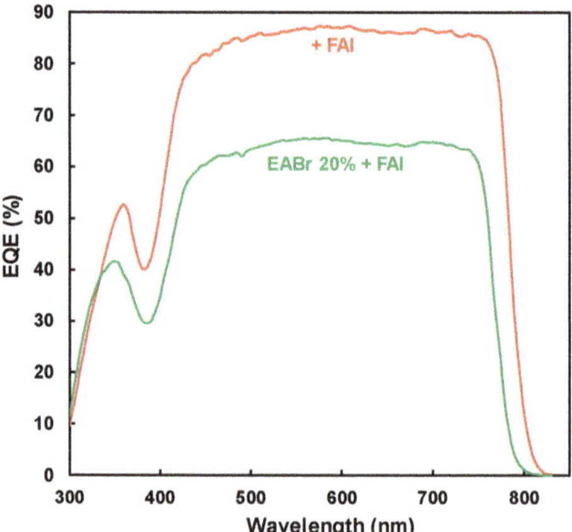

Figure 6. External quantum efficiency spectra of the fabricated solar cells.

X-ray diffraction (XRD) patterns of the fabricated cells added with EABr and FAI are shown in Figure 7a. Increases of (100) and (200) diffraction reflections are observed by adding FAI or EABr. In addition, only (100) and (200) peaks are observed, which indicates that the cells exhibited highly oriented (100) perovskite crystals by the air-blowing method [40].

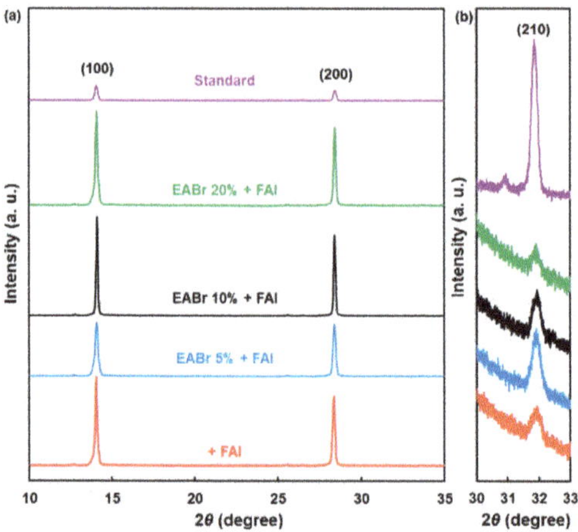

Figure 7. (**a**) X-ray diffraction (XRD) pattern of the present solar cells and (**b**) enlarged pattern of (**a**).

Microstructural parameters of the present perovskite compounds are listed in Table 4. The lattice constants of the FAI-added perovskites were higher compared with the standard $MAPbI_3$ material, whereas those of the EABr and FAI co-added perovskite decreased. Crystallite sizes were estimated from the (200) reflections, and they increased by the addition of FAI and EABr. The I_{100}/I_{210} intensity ratios of (100) reflections (I_{100}) to (210) reflections (I_{210}) were measured from the XRD data in Figure 7a,b, and the results are shown in Table 4. If the $CH_3NH_3PbI_3$ cubic perovskite particles are randomly oriented, then the I_{100}/I_{210} value should be 2.08 [35]. For the standard cell prepared in the present study, the I_{100}/I_{210} is 48, which means the (100) crystal surfaces of the cubic structures are strongly aligned in the solar cell. By the addition of FAI to the perovskite compounds, I_{100}/I_{210} was increased to 1694, and the I_{100}/I_{210} increased further to 1939 by adding EABr. This is 40 times higher than the I_{100}/I_{210} of the standard perovskite device.

Table 4. Microstructural parameters for the perovskite crystals. Preferred crystal orientations were indicated with ratios of 100 diffraction intensities (I_{100}) to 210 diffraction intensities (I_{210}).

Perovskites	Lattice Constant a (Å)	Crystallite Size D_{200} (Å)	Orientation I_{100}/I_{210}
Standard	6.274(1)	479	48
+FAI 20%	6.286(1)	647	1694
EABr 5% + FAI 20%	6.281(0)	528	460
EABr 10% + FAI 20%	6.283(1)	1506	1155
EABr 20% + FAI 20%	6.280(2)	830	1939

A schematic model showing molecular structures (MA, FA, and EA) and the lattice structure of the FAI and EABr added perovskites is shown in Figure 8a,b, respectively. The lattice constant a of 6.315 Å for a perovskite single crystal [35,45] is greater compared with the a of the perovskite compound in a cell configuration [46,47]. If the perovskite particles were synthesized and deposited on the mesoporous TiO_2 layer, some of the CH_3NH_2 molecules might be desorbed. Then, MA vacancies could be formed, and the lattice constant (6.274 Å) of $MAPbI_3$ is smaller than that of single crystal, as listed in Table 4. When FAI was added to the standard $MAPbI_3$, the FA would occupy the defects and MA sites, and the lattice constant increased to 6.286 Å, as shown in Figure 8b and Table 4. As the size of Br^- is fairly small compare with that of I^-, a values of the EABr-added crystals decreased to 6.280 Å compared with FAI-added perovskite crystals, as indicated by arrows in Figure 8b. Combination of the present EA/FA with other molecules [15,48] and alkali metals [21,49] might also be effective for the stabilization of the perovskite compounds.

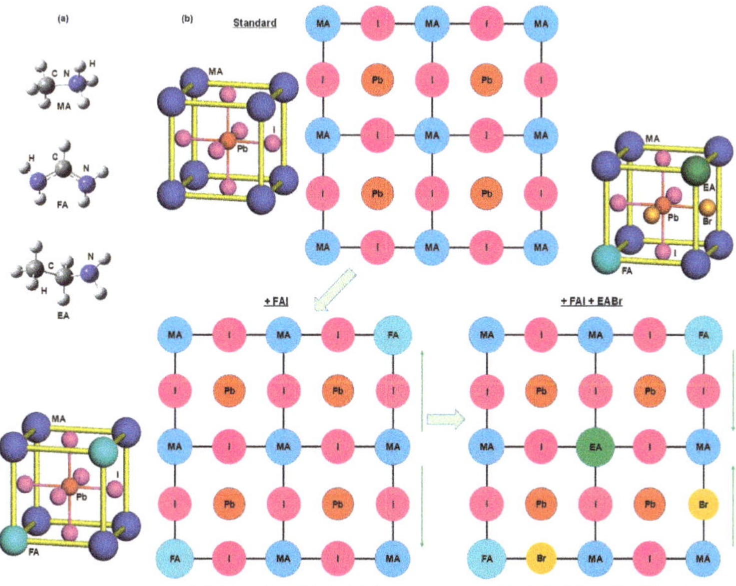

Figure 8. Structures of (**a**) methylammonium (MA), formamidinium (FA), ethylammonium (EA) and (**b**) the present perovskite compounds.

4. Conclusions

Solar cells using perovskite in the photoactive layer were produced by using spin-coating technique in ordinary air, and the influence on photovoltaic characteristics by adding EA and FA to the perovskite phase was investigated. From the results of J–V characteristics, the addition of EA and FA improved V_{OC} and FF, leading to an improvement in photoelectric conversion efficiency. Devices with EA and FA added maintained photoelectric conversion efficiencies even after 4 weeks compared to that of the standard device. Optical microscope results showed surface improvement, and X-ray diffraction results showed FA and EA substitution at MA position of the perovskite. By substituting FA and EA, which have larger ionic radii than MA, the perovskite structures would have more stable cubic structures with higher stability.

Author Contributions: Conceptualization, K.N. and T.O.; Methodology, K.N., T.K., N.U., and A.S.; Formal Analysis, K.N. and T.O.; Investigation, K.N.; Data Curation, K.N. and T.O.; Writing-Original Draft Preparation, K.N. and T.O.; Writing-Review & Editing, T.O.; Visualization, K.N. and T.O.; Supervision, T.O.; Project Administration, T.O.; Funding Acquisition, T.O. All authors have read and agreed to the published version of the manuscript.

Funding: This research was partly funded by the Super Cluster Program of the Japan Science and Technology Agency (JST).

Conflicts of Interest: The authors declare no conflict of interest.

References

1. Green, M.A.; Ho-Baillie, A.; Snaith, H.J. The emergence of perovskite solar cells. *Nat. Photon* **2014**, *8*, 506–514. [CrossRef]

2. Chen, Q.; Marco, N.D.; Yang, Y.M.; Song, T.B.; Chen, C.C.; Zhao, H.; Hong, Z.; Zhou, H.; Yang, Y. Under the spotlight: The organic—Inorganic hybrid halide perovskite for optoelectronic applications. *Nanotoday* **2015**, *10*, 355–396. [CrossRef]

3. Chen, Y.; He, M.; Peng, J.; Sun, Y.; Liang, Z. Structure and growth control of organic-inorganic halide perovskites for optoelectronics: From polycrystalline films to single crystals. *Adv. Sci.* **2016**, *3*, 1500392. [CrossRef] [PubMed]

4. Saliba, M.; Correa-Baena, J.P.; Wolff, C.M.; Stolterfoht, M.; Phung, N.; Albrecht, S.; Neher, D.; Ababe, A. How to make over 20% efficient perovskite solar cells in regular (n–i–p) and inverted (p–i–n) architectures. *Chem. Mater.* **2018**, *30*, 4193–4201. [CrossRef]

5. Wu, C.; Chen, K.; Guo, D.Y.; Wang, S.L.; Li, P.G. Cations substitution tuning phase stability in hybrid perovskite single crystals by strain relaxation. *RSC Adv.* **2018**, *8*, 2900–2905. [CrossRef]

6. Zhao, Y.; Ye, Q.; Chu, Z.; Gao, F.; Zhang, X.; You, J. Recent progress in high-efficiency planar-structure perovskite solar cells. *Energy Environ. Mater.* **2019**, *2*, 93–106. [CrossRef]

7. Weller, M.T.; Weber, O.J.; Frost, J.M.; Walsh, A. Cubic perovskite structure of black Formamidinium Lead Iodide, α-[HC(NH$_2$)$_2$]PbI$_3$, at 298 K. *J. Phys. Chem. Lett.* **2015**, *6*, 3209–3212. [CrossRef]

8. Zhao, Y.; Tan, H.; Yuan, H.; Yang, Z.; Fan, J.Z.; Kim, J.; Voznyy, O.; Gong, X.; Quan, L.N.; Tan, C.S.; et al. Perovskite seeding growth of formamidinium-lead-iodide-based perovskites for efficient and stable solar cells. *Nature* **2018**, *9*, 1607. [CrossRef]

9. Yang, W.S.; Park, B.W.; Jung, E.H.; Jeon, N.J.; Kim, Y.C.; Lee, D.U.; Shin, S.S.; Seo, J.; Kim, E.K.; Noh, J.H.; et al. Iodide management in formamidinium-lead-halide-based perovskite layers for efficient solar cells. *Science* **2017**, *356*, 1376–1379. [CrossRef]

10. Chen, Z.; Zheng, X.; Yao, F.; Ma, J.; Tao, C.; Fang, G. Methylammonium, formamidinium and ethylenediamine mixed triple-cation perovskite solar cells with high efficiency and remarkable stability. *J. Mater. Chem. A* **2018**, *6*, 17625–17632. [CrossRef]

11. Shao, S.; Dong, J.; Duim, H.; Brink, G.H.T.; Black, G.R.; Portale, G.; Loi, M.A. Enhancing the crystallinity and perfecting the orientation of formamidinium tin iodide for highly efficient Sn-based perovskite solar cells. *Nano Energy* **2019**, *60*, 810–816. [CrossRef]

12. Tavakoli, M.M.; Yadav, P.; Prochowicz, D.; Sponseller, M.; Osherov, A.; Bulovic, V.; Kong, J. Controllable perovskite crystallization via antisolvent technique using chloride additives for highly efficient planar perovskite solar cells. *Adv. Energy Mater.* **2019**, *9*, 1803587. [CrossRef]

13. Lu, Y.A.; Chang, T.H.; Wu, S.H.; Liu, C.C.; Lai, K.W.; Chang, Y.C.; Lu, H.C.; Chu, C.W.; Ho, K.C. Coral-like perovskite nanostructures for enhanced light-harvesting and accelerated charge extraction in perovskite solar cells. *Nano Energy* **2019**, *58*, 138–146. [CrossRef]

14. Alharbi, E.A.; Alyamani, A.Y.; Kubicki, D.J.; Uhl, A.R.; Walder, B.J.; Alanazi, A.Q.; Luo, J.; Burgos-Caminal, A.; Albadri, A.; Albrithen, H.; et al. Atomic-level passivation mechanism of ammonium salts enabling highly efficient perovskite solar cells. *Nat. Commun.* **2019**, *10*, 3008. [CrossRef] [PubMed]

15. Kishimoto, T.; Suzuki, A.; Ueoka, N.; Oku, T. Effects of guanidinium addition to $CH_3NH_3PbI_{3-x}Cl_x$ perovskite photovoltaic devices. *J. Ceram. Soc. Jpn.* **2019**, *127*, 491–497. [CrossRef]

16. Hu, Y.; Ayguler, M.F.; Petrus, M.L.; Bein, T.; Docampo, P. Impact of rubidium and cesium cations on the moisture stability of multiple-cation mixed-halide perovskites. *ACS Energy Lett.* **2017**, *2*, 2212–2218. [CrossRef]

17. Hu, Y.; Hutter, E.M.; Rieder, P.; Grill, I.; Hanisch, J.; Ayguler, M.F.; Hufnagel, A.G.; Handloser, M.; Bein, T.; Hartschuh, A.; et al. Understanding the role of cesium and rubidium additives in perovskite solar cells: Trap states, charge transport, and recombination. *Adv. Energy Mater.* **2018**, *8*, 1703057. [CrossRef]

18. Zhao, W.; Yao, Z.; Yu, F.; Yang, D.; Liu, S.F. Alkali metal doping for improved $CH_3NH_3PbI_3$ perovskite solar cells. *Adv. Sci.* **2018**, *5*, 1700131. [CrossRef]

19. Ling, T.; Zou, X.; Cheng, J.; Yang, Y.; Ren, H.; Chen, D. Modulating surface morphology related to crystallization speed of perovskite grain and semiconductor properties of optical absorber layer under controlled doping of potassium ions for solar cells. *Materials* **2018**, *11*, 1605. [CrossRef]

20. Machiba, H.; Oku, T.; Kishimoto, T.; Ueoka, N.; Suzuki, A. Fabrication and evaluation of K-doped $MA_{0.8}FA_{0.1}K_{0.1}PbI_3(Cl)$ perovskite solar cells. *Chem. Phys. Lett.* **2019**, *730*, 117–123. [CrossRef]

21. Ueoka, N.; Oku, T.; Suzuki, A. Additive effects of alkali metals on Cu-modified $CH_3NH_3PbI_{3-\delta}Cl_\delta$ photovoltaic devices. *RSC Adv.* **2019**, *9*, 24231–24240. [CrossRef]

22. Hsu, H.L.; Chang, C.C.; Chen, C.P.; Jiang, B.H.; Jeng, R.J.; Cheng, C.H. High-performance and high-durability perovskite photovoltaic devices prepared using ethylammonium iodide as an additive. *J. Mater. Chem. A* **2015**, *3*, 9271–9277. [CrossRef]

23. Zheng, H.; Liu, G.; Chen, X.; Zhang, B.; Alsaedi, A.; Hayat, T.; Pan, X.; Dai, S. High-performance mixed-dimensional perovskite solar cells with enhanced stability against humidity, heat and UV light. *J. Mater. Chem. A* **2018**, *6*, 20233–20241. [CrossRef]

24. Wang, Y.; Zhang, T.; Li, G.; Xu, F.; Li, Y.; Yang, Y.; Zhao, Y. A mixed-cation lead iodide $MA_{1-x}EA_xPbI_3$ absorber for perovskite solar cells. *J. Energy Chem.* **2018**, *27*, 215–218. [CrossRef]

25. Liu, D.; Li, Q.; Wu, K. Ethylammonium as an alternative cation for efficient perovskite solar cells from first-principles calculations. *RSC Adv.* **2019**, *9*, 7356–7361. [CrossRef]

26. Dhar, A.; Dey, A.; Maiti, P.; Paul, P.K.; Roy, S.; Paul, S.; Vekariya, R.L. Fabrication and characterization of next generation nano-structured organo-lead halide-based perovskite solar cell. *Ionics* **2018**, *24*, 1227–1233. [CrossRef]

27. Arkan, F.; Mohammad, I. Computational modeling of the photovoltaic activities in $EABX_3$ (EA = ethylammonium, B = Pb, Sn, Ge, X = Cl, Br, I) perovskite solar cells. *Comput. Mater. Sci.* **2018**, *152*, 324–330. [CrossRef]

28. Zhang, F.; Cong, J.; Li, Y.; Bergstrand, J.; Liu, H.; Cai, B.; Hajian, A.; Yao, Z.; Wang, L.; Hao, Y.; et al. A facile route to grain morphology controllable perovskite thin films towards highly efficient perovskite solar cells. *Nano Energy* **2018**, *53*, 405–414. [CrossRef]

29. Liu, F.; Dong, Q.; Wong, M.K.; Djiurisic, A.B.; Ng, A.; Ren, Z.; Shen, Q.; Surya, C.; Chan, W.K.; Wang, J.; et al. Perovskite solar cells: Is excess PbI_2 beneficial for perovskite solar cell performance? *Adv. Energy Mater.* **2016**, *6*, 1502206. [CrossRef]

30. Ueoka, N.; Oku, T. Stability characterization of PbI_2-added $CH_3NH_3PbI_{3-x}Cl_x$ photovoltaic devices. *ACS Appl. Mater. Interfaces* **2018**, *10*, 44443–44451. [CrossRef]

31. Hoefler, S.F.; Trimmel, G.; Rath, T. Progress on lead-free metal halide perovskites for photovoltaic applications: A review. *Mon. Chem.* **2017**, *148*, 795–826. [CrossRef]

32. Xu, F.; Zhang, T.; Li, G.; Zhao, Y. Mixed cation hybrid lead halide perovskites with enhanced performance and stability. *J. Mater. Chem. A* **2017**, *5*, 11450–11461. [CrossRef]

33. Sampson, M.D.; Park, J.S.; Schaller, R.D.; Chan, M.K.Y.; Martinson, A.B.F. Transition metal-substituted lead halide perovskite absorbers. *J. Mater. Chem. A* **2017**, *5*, 3578–3588. [CrossRef]

34. Tanaka, H.; Oku, T.; Ueoka, N. Structural stabilities of organic–inorganic perovskite crystals. *Jpn. J. Appl. Phys.* **2018**, *57*, 08RE12. [CrossRef]

35. Oku, T. Crystal structures of perovskite halide compounds used for solar cells. *Rev. Adv. Mater. Sci.* **2020**, *59*. in press. [CrossRef]

36. Shannon, R.D. Revised effective ionic radii and systematic studies of interatomic distances in halides and chalcogenides. *Acta Cryst. A* **1976**, *32*, 751–767. [CrossRef]

37. Taguchi, M.; Suzuki, A.; Oku, T.; Fukunishi, S.; Minami, S.; Okita, M. Effects of decaphenylcyclopentasilane addition on photovoltaic properties of perovskite solar cells. *Coatings* **2018**, *8*, 461. [CrossRef]

38. Oku, T.; Nomura, J.; Suzuki, A.; Tanaka, H.; Fukunishi, S.; Minami, S.; Tsukada, S. Fabrication and characterization of $CH_3NH_3PbI_3$ perovskite solar cells added with polysilanes. *Int. J. Photoenergy* **2018**, *1155*, 1–7. [CrossRef]

39. Taguchi, M.; Suzuki, A.; Oku, T.; Ueoka, N.; Minami, S.; Okita, M. Effects of annealing temperature on decaphenylcyclopentasilane-inserted $CH_3NH_3PbI_3$ perovskite solar cells. *Chem. Phys. Lett.* **2019**, *737*, 136822. [CrossRef]

40. Oku, T.; Ohishi, Y.; Ueoka, N. Highly (100)-oriented $CH_3NH_3PbI_3$ (Cl) perovskite solar cells prepared with NH_4Cl using an air blow method. *RSC Adv.* **2018**, *8*, 10389–10395. [CrossRef]

41. Oku, T.; Ohishi, Y. Effects of annealing on $CH_3NH_3PbI_3$ (Cl) perovskite photovoltaic devices. *J. Ceram. Soc. Jpn.* **2018**, *126*, 56–60. [CrossRef]

42. Oku, T.; Ueoka, N.; Suzuki, K.; Suzuki, A.; Yamada, M.; Sakamoto, H.; Minami, S.; Fukunishi, S.; Kohno, K.; Miyauchi, S. Fabrication and characterization of perovskite photovoltaic devices with TiO_2 nanoparticle layers. *AIP Conf. Proc.* **2017**, *1807*, 020014-1–020014-7.

43. Ueoka, N.; Oku, T.; Suzuki, A.; Sakamoto, H.; Yamada, M.; Minami, S.; Miyauchi, S. Fabrication and characterization of $CH_3NH_3(Cs)Pb(Sn)I_3(Cl)$ perovskite solar cells with TiO_2 nanoparticle layers. *Jpn. J. Appl. Phys.* **2018**, *57*, 02CE03-1–02CE03-7. [CrossRef]

44. Ueoka, N.; Oku, T.; Tanaka, H.; Suzuki, A.; Sakamoto, H.; Yamada, M.; Minami, S.; Miyauchi, S.; Tsukada, S. Effects of PbI_2 addition and TiO_2 electron transport layers for perovskite solar cells. *Jpn. J. Appl. Phys.* **2018**, *57*, 08RE05-1–08RE05-7. [CrossRef]

45. Ren, Y.; Oswald, I.W.H.; Wang, X.; McCandless, G.T.; Chan, J.Y. Orientation of organic cations in hybrid inorganic–organic perovskite $CH_3NH_3PbI_3$ from subatomic resolution single crystal neutron diffraction structural studies. *Cryst. Growth Des.* **2016**, *16*, 2945–2951. [CrossRef]

46. Oku, T.; Zushi, M.; Imanishi, Y.; Suzuki, A.; Suzuki, K. Microstructures and photovoltaic properties of perovskite-type $CH_3NH_3PbI_3$ compounds. *Appl. Phys. Express* **2014**, *7*, 121601. [CrossRef]

47. Oku, T.; Ohishi, Y.; Suzuki, A. Effects of antimony addition to perovskite-type $CH_3NH_3PbI_3$ photovoltaic devices. *Chem. Lett.* **2016**, *45*, 134–136. [CrossRef]

48. Jodlowski, A.; Roldán-Carmona, C.; Grancini, G.; Salado, M.; Ralaiarisoa, M.; Ahmad, S.; Koch, N.; Camacho, L.; Miguel, G.; Nazeeruddin, M. Large guanidinium cation mixed with methylammonium in lead iodide perovskites for 19% efficient solar cells. *Nat. Energy* **2017**, *2*, 972–979. [CrossRef]

49. Jalebi, M.A.; Garmaroudi, Z.A.; Pearson, A.J.; Divitini, G.; Cacovich, S.; Philippe, B.; Rensmo, H.; Ducati, C.; Friend, R.H.; Stranks, S.D. Potassium- and rubidium-passivated alloyed perovskite films: Optoelectronic properties and moisture stability. *ACS Energy Lett.* **2018**, *3*, 2671–2678. [CrossRef]

Article

Effect of Thioacetamide Concentration on the Preparation of Single-Phase SnS and SnS$_2$ Thin Films for Optoelectronic Applications

Sreedevi Gedi [1,†], Vasudeva Reddy Minnam Reddy [1,†], Salh Alhammadi [1], Doohyung Moon [1], Yeongju Seo [1], Tulasi Ramakrishna Reddy Kotte [2], Chinho Park [1] and Woo Kyoung Kim [1,*]

[1] School of Chemical Engineering, Yeungnam University, 280 Daehak-ro, Gyeongsan 38541, Korea; drsrvi9@gmail.com (S.G.); drmvasudr9@gmail.com (V.R.M.R.); salehalhammadi.1987@gmail.com (S.A.); engud011@naver.com (D.M.); tjdudwn01@naver.com (Y.S.); chpark@ynu.ac.kr (C.P.)

[2] Solar Energy Laboratory, Department of Physics, Sri Venkateswasra University, Tirupati 517 502, India; ktrkreddy@gmail.com

* Correspondence: wkim@ynu.ac.kr; Tel.: +82-53-810-2514; Fax: +82-53-810-4631

† These authors contributed equally to this work.

Received: 5 September 2019; Accepted: 27 September 2019; Published: 30 September 2019

Abstract: Eco-friendly tin sulfide (SnS) thin films were deposited by chemical solution process using varying concentrations of a sulfur precursor (thioacetamide, 0.50–0.75 M). Optimized thioacetamide concentrations of 0.6 and 0.7 M were obtained for the preparation of single-phase SnS and SnS$_2$ films for photovoltaic absorbers and buffers, respectively. The as-deposited SnS and SnS$_2$ thin films were uniform and pinhole-free without any major cracks and satisfactorily adhered to the substrate; they appeared in dark-brown and orange colors, respectively. Thin-film studies (compositional, structural, optical, and electrical) revealed that the as-prepared SnS and SnS$_2$ films were polycrystalline in nature; exhibited orthorhombic and hexagonal crystal structures with (111) and (001) peaks as the preferred orientation; had optimal band gaps of 1.28 and 2.92 eV; and exhibited p- and n-type electrical conductivity, respectively. This study presents a step towards the growth of SnS and SnS$_2$ binary compounds for a clean and economical power source.

Keywords: tin monosulfide; tin disulfide; chemical solution process; absorber; buffer; solar cells; renewable energy

---

1. Introduction

In recent years, inorganic semiconductors have been employed to generate clean energy in various optoelectronic and electrochemical fields, mainly because of their high stability under atmospheric conditions [1–5]. Among these, groups IV–VI tin-based binary sulfides such as tin monosulfide (SnS) and tin disulfide (SnS$_2$) have drawn much attention because of their abundance, low cost, non-toxicity, high catalytic activity, material diversity, structural multiformity, and excellent electrical conductivity [6]. SnS crystallizes in an orthorhombic structure with the Pnma space group [7] and exhibits excellent optoelectrical properties such as a strong absorption coefficient (> 10^5 cm^{-1}) with direct optical band gap energy ranging from 1.16 to 1.79 eV [8,9] and p-type conductivity with a carrier concentration of the order of 10^{11}–10^{18} cm^{-3} [10], which are suitable for thin-film photovoltaic absorbers. In addition, maintenance of stoichiometry is simpler compared with the other conventional light absorbers [11–15]. Another tin-based binary sulfide, SnS$_2$, adopts a hexagonal-layered structure with the P3ml space group, similar to the structure of the CdI$_2$ system [16]. Moreover, SnS$_2$ has a relatively high optical band gap in the range of 2.04–3.30 eV [17] with direct transitions and exhibits n-type conductivity. These promising characteristics make SnS a suitable absorber candidate material for thin-film solar cells [1]; moreover, SnS is presumably expected to be a better alternative to the already developed technologies such as Cu(In,Ga)Se$_2$ (23.35%) [18] and CdTe (22.1%) [19], which are limited by their toxicity, scarcity, and exorbitant cost. Similarly, the favorable electronic characteristics, excellent structural flexibility, wider spectral response, and better thermal stability of SnS$_2$ make it a competitive nontoxic substitute for the conventional CdS [14] and Zn(O,S) [20] buffers in thin-film solar cells [2]. Furthermore, both SnS and SnS$_2$ have also been applied in photodetectors [21,22], photocatalysis [23], water splitting [24], gas sensors [25,26], biosensors [27], field-effect transistors [28,29], lithium-sodium ion batteries [30,31], thermoelectrics [32], electrochemical capacitors [33], and supercapacitors [34].

Various approaches including physical and chemical techniques have been explored to deposit both SnS and SnS$_2$ thin films. Among these techniques, the chemical solution process (CSP) or chemical bath deposition (CBD) offers more inexpensive and flexible depositions. The preparation of single-phase SnS and SnS$_2$ films by CSP depends on the release and reaction rate of sulfur ions in the reaction solution, which can be controlled by the selection of a suitable sulfur precursor at a concentration, which affects not only the formation of the phase but also the growth rate, thickness, and other physical properties of the product films. Therefore, optimizing the concentration of the sulfur precursor is highly preferable to obtain high-quality SnS and SnS$_2$ films for the desired applications. Previous research suggests that thioacetamide (TA) and sodium thiosulfate have been mainly used as sulfur precursors since 1987 (Figure 1). Of these, thioacetamide is more preferable because it works under both acidic and alkaline bath conditions.

Therefore, a set of SnS thin films were deposited by CSP using varying thioacetamide concentrations, while the other growth parameters were maintained constant. The influence of thioacetamide concentration on the growth and physical properties of the thin films are discussed in the subsequent sections. Thus, its concentration was optimized for the preparation of pure SnS and SnS$_2$ films for wide applications.

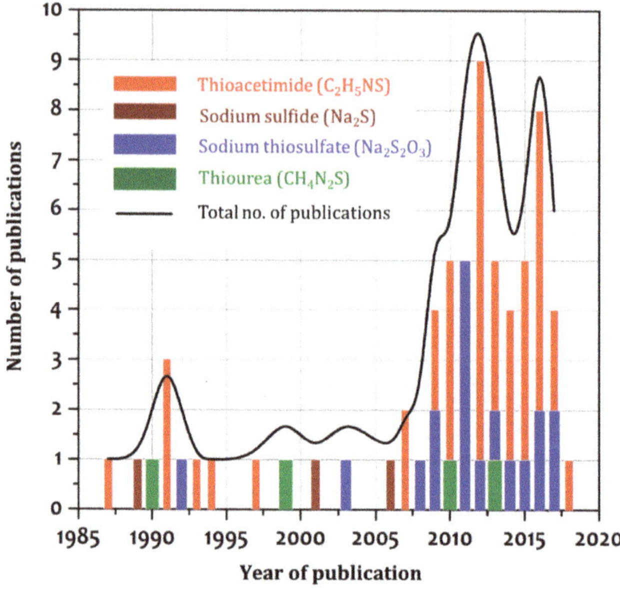

Figure 1. The number of chemical bath deposition (CBD) SnS reports on the usage of different sulfur precursors from 1980 to the present.

2. Experimental Section

The deposition procedure for SnS and SnS$_2$ films by CSP is schematically presented in Figure 2, and the process has been described in detail in our previous reports [35,36]. Analytical-grade tin chloride (SnCl$_2$.2H$_2$O, 0.1 M) and tartaric acid (C$_4$H$_6$O$_6$, 1.2 M) were used as a tin source and complexing agent, respectively, and 10 mL of thioacetamide (C$_2$H$_5$NS, 0.50-0.75 M) was used as the source material for the S ions. The as-prepared thin films were cleaned using deionized (DI) water and dried in a vacuum oven. By the following equations, the kinetics of SnS and SnS$_2$ films can be understood.

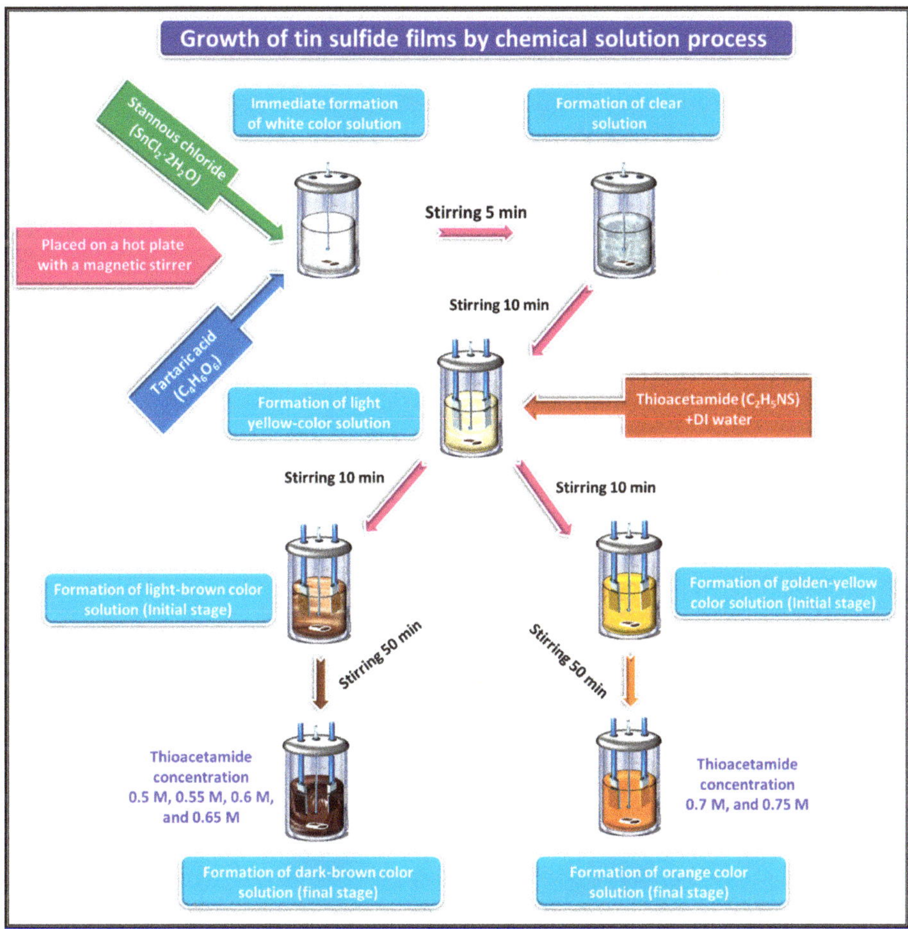

Figure 2. Schematic representation of the growth of tin sulfide films.

In an aqueous tin chloride solution, the complexation of tartaric acid is

$$SnCl_2 \cdot 2H_2O + L \Leftrightarrow Sn(C_4H_6O_6)^{2+} + 2Cl^- + 2H_2O.$$

The free Sn^{2+} ions are slowly released by the dissociation of tin tartrate ions (Sn(C$_4$H$_6$O$_6$)$^{2+}$) in a controlled way, as below:

$$Sn(C_4H_6O_6)^{2+} \rightleftharpoons Sn^{2+} + C_4H_6O_6.$$

The hydrolysis of thioacetamide can produce S^{2-} ions by

$$CH_3CSNH_2 + H_2O \rightleftharpoons CH_3CONH_2 + H_2S,$$

$$H_2S + H_2O \rightleftharpoons H_3O^+ + HS^-,$$

$$HS^- \rightleftharpoons H^+ + S^{2-},$$

$$HS^- + OH^- \rightleftharpoons H_2O + S^{2-}.$$

When the thioacetamide concentration is changed, multiphase or other single-phase films can be formed by the following reactions:

$$Sn^{2+} + S^{2-} \rightarrow SnS,$$

$$3SnS + S \rightarrow SnS + Sn_2S_3,$$

$$Sn_2S_3 + S \rightarrow 2SnS_2,$$

The compositional, structural, optical, and electrical characteristics of the as-deposited films were analyzed using an X-ray photoelectron spectrometer (Waltham, MA, USA) (XPS: A VG Multilab 2000; Al Kα radiation = 1486.6 eV), X-ray diffractometer (Almelo, Overijssel, Netherlands) (XRD: PANalytical X'Pert PRO MPD, Malvern Panalytical with Cu Kα radiation, λ = 1.5406 Å), UV-Vis-NIR spectrometer (Santa Clara, CA, USA) (Aglient), and Hall measurement system (Anyang-city, Gyeonggi-do, South Korea) (ECOPIA HMS-3000VER), respectively.

3. Results and Discussion

All the deposited tin sulfide films were confirmed to be satisfactorily adherent. The films appeared in dark-brown and orange shades at TA concentration ranges of 0.50–0.65 M and 0.70–0.75 M, respectively. The effect of TA concentration on the elemental composition and chemical states of elements was investigated by XPS. The XPS survey spectrum indicated that the deposited films were mainly composed of the Sn 3d doublet along with the S 2p peaks, representing the existence of the elements Sn and S. A high-resolution scan of the Sn 3d and S 2p peaks is shown in Figure 3a,b, respectively.

The Sn 3d doublet in the films grown with lower TA concentrations (0.5–0.6 M) showed two strong peaks at about 485.6 and 494.1 eV that were related to Sn $3d_{5/2}$ and Sn $3d_{3/2}$, respectively, which are the characteristics of Sn^{2+} in SnS. The Sn $3d_{5/2}$ and Sn $3d_{3/2}$ peaks were observed to expand at a TA concentration of 0.65 M, and the deconvolution of these peaks reveals the existence of two humps in each peak, linked to the elevated intensity for Sn^{2+} and low intensity for Sn^{4+}, indicating the coexistence of secondary Sn_2S_3 along with SnS. As the TA concentration was increased further (0.7 M), the spectra showed the binding energies for Sn $3d_{5/2}$ and Sn $3d_{3/2}$ as 486.7 (487.2) and 494.9 (495.1) eV, respectively [37,38], revealing the presence of Sn^{4+} produced in the SnS_2 phase. Beyond this concentration (0.75 M), the deconvolution of the broadened Sn 3d doublets suggests the presence of low intense Sn^{2+} and high intense Sn^{4+}, confirming the formation of the Sn_2S_3 secondary phase along with SnS_2. The high-resolution XPS spectrum of S 2p for the as-prepared films exhibited single peaks related to S^{2-}–Sn^{2+} (161.3 eV) and S–Sn^{4+} (162.3 eV) bonds at TA concentrations of 0.5–0.6 and 0.7 M, which are characteristic of single-phase SnS and SnS_2 films, respectively [35]. The remaining core-level S 2p spectra (for TA = 0.65 and 0.75 M) showed two deconvoluted peaks related to the binding energies of both S^{2-}–Sn^{2+} (161.3 eV) and S–Sn^{4+} (162.3 eV) bonding, suggesting the presence of the Sn_2S_3 secondary phase. The Sn/S atomic ratio varied between 1.12 to 0.53 with the change in TA concentration. The lower Sn/S atomic ratio indicated that S is more dominant at higher TA values. This excess of sulfur probably led to the formation of the binary phases, SnS_2 and Sn_2S_3 along with the SnS phase, which were further observed in the XRD studies. The film grown at TA concentrations of 0.6 M and 0.7 M exhibited Sn/S atomic ratios of 0.98 and 0.51, indicating that the films have a clear stoichiometry for the formation of the pure SnS and SnS_2 phases, respectively.

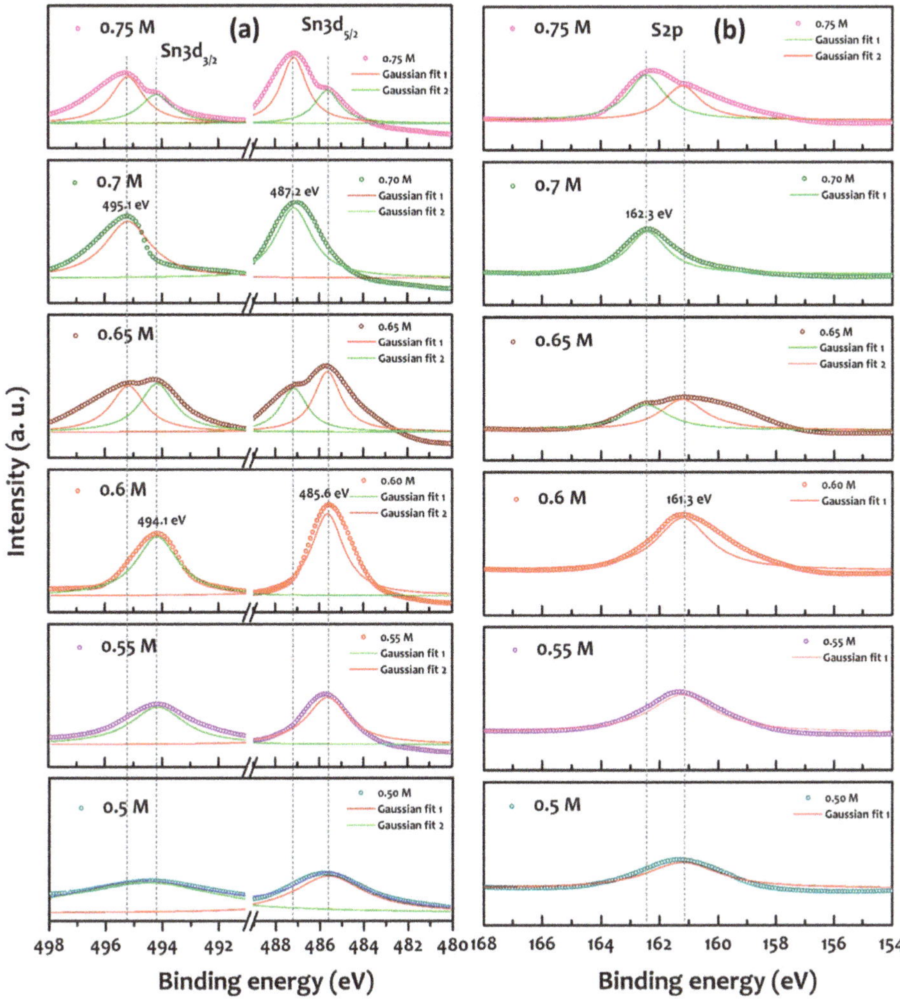

Figure 3. High-resolution X-ray photoelectron spectrometer (XPS) scan of (**a**) Sn 3d and (**b**) S 2p core levels.

In general, the structural properties of the thin films were observed to be highly influenced by the compositional variations in the film. In this study, the structural behavior of SnS films grown using different TA concentrations was analyzed using XRD. The diffraction patterns (Figure 4) confirmed the polycrystalline nature of the as-prepared films; some of these showed various binary phases, depending upon the TA concentration: the films deposited in the TA concentration range of 0.5−0.6 M exhibited a strong peak at $2\theta = 31.6°$, which is related to the (111) plane, along with other minor peaks at around $2\theta = 27.1°$, $38.8°$, and $51.1°$, which were assigned to the diffraction from the (101), (131), and (112) planes of orthorhombic SnS (JCPDS Card No. 39-0354) [39], respectively. Moreover, the crystalline state improved with an increase in the TA concentration. As the TA concentration was increased to 0.55 M, the film exhibited additional planes of (130) and (260) related to the Sn_2S_3 binary phase (JCPDS Card No. 14-0619) [40]; as the TA concentration was further increased to 0.7 M, the film exhibited different diffraction planes of hexagonal SnS_2 (JCPDS Card No. 23-0677) [41] with (001) reflection at 15.02° as the preferred orientation; and beyond this TA concentration (0.75 M), the film additionally exhibited (130), (260), and (540) planes of the secondary Sn_2S_3 phase along with SnS_2. Therefore, the films deposited under TA concentrations of 0.6 and 0.7 M exhibited good crystalline and pure orthorhombic SnS (111) and hexagonal SnS_2 (001) phases, respectively. Thus, these results are observed to agree well with the XPS observations.

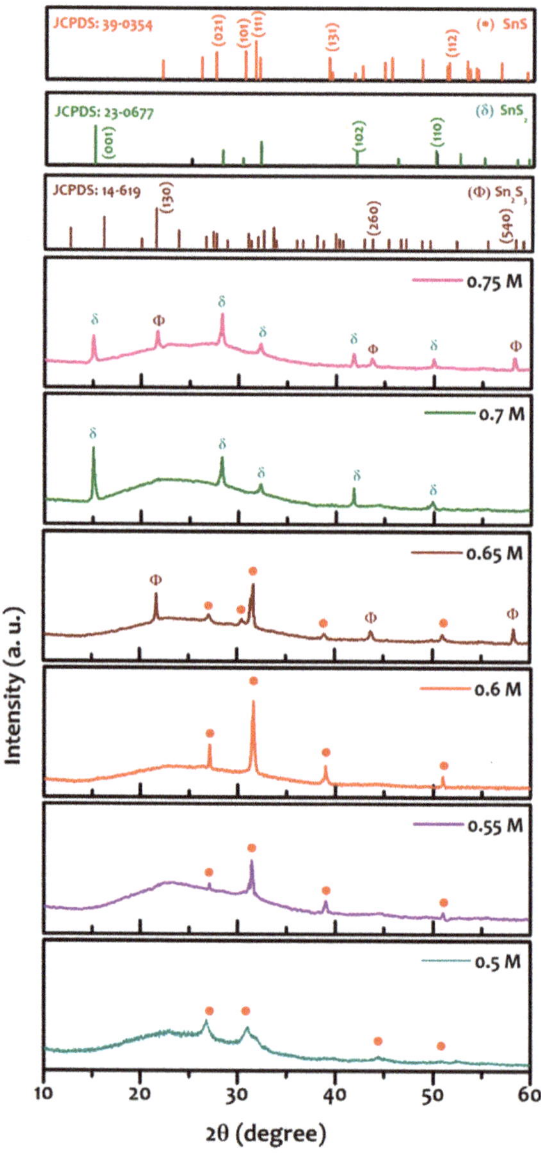

Figure 4. XRD patterns of the films deposited at different TA concentrations.

The average crystallite size, D, of the deposited films was estimated using the Debye–Scherrer formula [42] using the strong (1 1 1) and (0 0 1) peaks of the deposited films with a TA concentration in the range of 0.5–0.65 M and 0.7–0.75 M, respectively. It was found that the crystal size varied from

15 to 29 nm with the change of TA concentration from 0.5 to 0.65 M. In addition, a D of 26 nm was obtained for the film deposited at a concentration of 0.7 M, and it was decreased to 24 nm for a further increase in TA concentration to 0.75 M. Therefore, the D of the SnS (TA = 0.6 M) and SnS_2 (0.7 M) films was obtained as 29 nm and 26 nm, respectively. The surface morphology and surface average grain size of both SnS (TA = 0.6 M) and SnS_2 (0.7 M) films were obtained from SEM images (Figure 5). Both the films showed compact morphology without pin-holes and cracks. The SnS film showed almost elliptical-shaped grains with an average grain size of ~295 nm, whereas the SnS_2 film showed shuttle-shaped grains of nearly 272 nm in size.

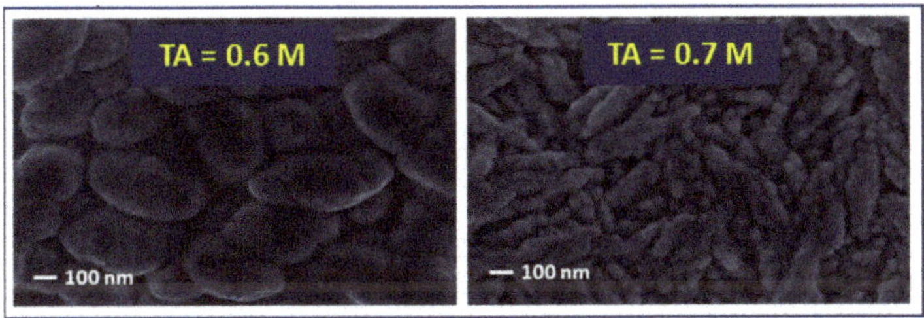

Figure 5. SEM surface images of both SnS and SnS_2 films deposited at thioacetamide (TA) concentrations of 0.6 and 0.7 M, respectively.

The optical studies of the films deposited at different TA concentrations were carried out using the UV-Vis-NIR photo spectrometer. The films deposited at TA concentrations in the range of 0.5–0.6 and 0.7 M exhibited a sharp absorption edge in the absorption spectra, whereas at TA concentrations of 0.65 and 0.75 M, the films exhibited a nonlinear fall in the absorption edge, indicating the potential presence of the secondary phase of Sn_2S_3 along with either the SnS or SnS_2 phase. The band gap of the prepared films was estimated from the Tauc plots ($(\alpha h\nu)^2$ vs. $h\nu$) [43], as shown in Figure 6. The films deposited at concentrations of 0.6 and 0.7 M are observed to exhibit band gaps of 1.28 and 2.92 eV, respectively, corresponding to the energy gaps of the SnS and SnS_2 phases. Furthermore, the films prepared at a 0.65 M concentration exhibited two different values of band gap—specifically, ~0.94 and ~1.34 eV—that are attributed to the presence of the Sn_2S_3 secondary phase along with SnS; on the other hand, at a concentration of 0.75 M, the films exhibited band gaps of ~1.11 and ~2.53 eV, corresponding to the presence of the Sn_2S_3 secondary phase and SnS_2 phase, respectively.

Figure 7 shows the variation of the refractive index (n) and extinction coefficient (k) as a function of the TA concentration. The evaluated n value is observed to first increase and then decrease with increasing TA concentrations up to 2 M, whereas the calculated k value exhibits a reverse variation trend. The films grown at TA concentrations of 0.6 and 0.7 M exhibit n values of 3.24 and 2.63 and k values of 0.13 and 0.61, respectively. The obtained n and k values in the present study are in good agreement with the reported values [2,44].

Electrical parameters such as conductivity type, resistivity, mobility, and carrier density of the films deposited at various TA concentrations were determined using Hall effect studies. The positive sign of the Hall coefficient for the films prepared at TA concentrations in the range of 0.50–0.65 M indicates p-type electrical conductivity, whereas the negative sign for the films deposited at TA concentrations of 0.7 and 0.75 M suggests n-type electrical conductivity. Furthermore, the films deposited at concentrations of 0.6 and 0.7 M exhibited carrier densities of 4.1×10^{15} and 7.2×10^{19} cm^{-3}; mobilities of 62.3 and 32.4 $cm^2V^{-1}s^{-1}$; and resistivities of 29.2 and 17.4 Ω-cm, respectively, which are related to the SnS and SnS_2 phases. These results are in agreement with the reported data [45,46].

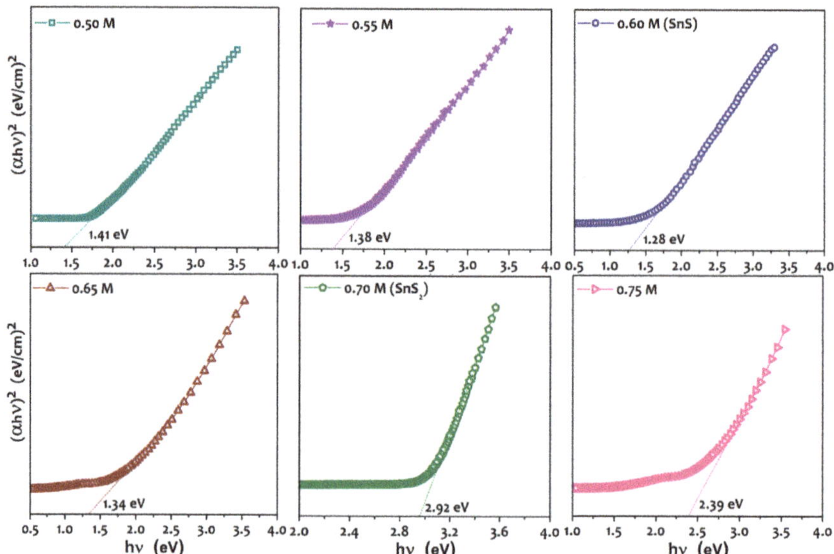

Figure 6. Tauc's plots of the films deposited at different TA concentrations.

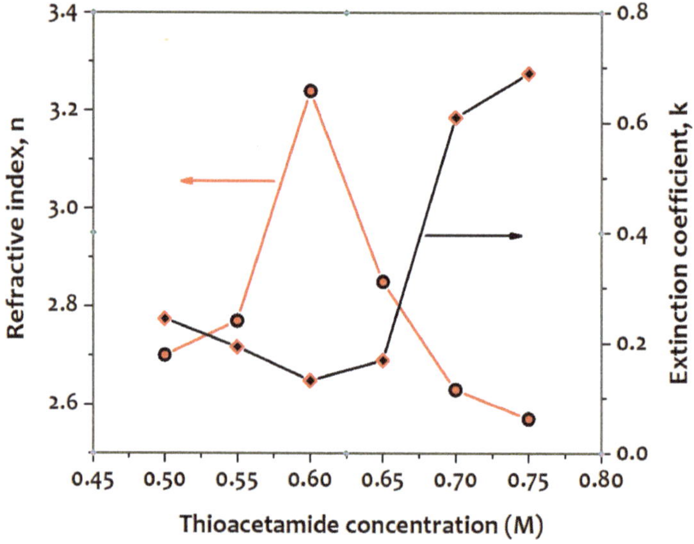

Figure 7. Variation of n and k with TA concentrations.

4. Conclusions

In this study, SnS thin films were prepared using chemical solution process by varying the concentration of TA (0.5–0.75 M), while the other growth parameters were maintained constant. The films prepared at TA concentrations of 0.6 and 0.7 M exhibited Sn/S atomic ratios of 0.98 and 0.51, confirming the formation of single SnS and SnS_2 phases, respectively. The XRD studies showed that all deposited films were polycrystalline in nature and exhibited various binary phases depending upon the TA concentration. The films deposited at a TA concentration of 0.6 M exhibited a pure orthorhombic SnS phase, whereas those at 0.7 M concentration exhibited a completely hexagonal SnS_2

phase. In addition, they showed the direct optical band gap of 1.28 and 2.92 eV with p- and n-type conductivities, which confirmed the suitability of these films not only for application as solar absorbers and buffers, respectively, but also for other applications.

Author Contributions: Conceptualization, T.R.R.K. and W.K.K.; Data curation, C.P. and W.K.K.; Formal analysis, S.G., V.R.M.R., S.A., D.M., and Y.S.; Funding acquisition, C.P.; Investigation, W.K.K.; Methodology, T.R.R.K. and C.P.; Supervision, W.K.K.; Writing – original draft, S.G., V.R.M.R., and W.K.K.

Funding: This research was funded by Priority Research Centers Program through the National Research Foundation of Korea (NRF) funded by the Ministry of Education (2014R1A6A1031189) and "Human Resources Program in Energy Technology" of the Korea Institute of Energy Technology Evaluation and Planning (KETEP), and granted financial resource from the Ministry of Trade, Industry and Energy, Republic of Korea (No. 20174030201760).

Conflicts of Interest: The authors declare no conflict of interest.

References

1. Minnam Reddy, V.R.; Gedi, S.; Pejjai, B.; Kotte, T.R.; Guillaum, Z.; Park, C. Influence of different substrates on the properties of sulfurized sns films. *Sci. Adv. Mater.* **2016**, *8*, 247–251. [CrossRef]
2. Gedi, S.; Minnam Reddy, V.R.; Pejjai, B.; Park, C.; Jeon, C.W.; Kotte, T.R. Studies on chemical bath deposited SnS 2 films for Cd-free thin film solar cells. *Ceram. Int.* **2017**, *43*, 3713–3719. [CrossRef]
3. Jung, H.-R.; Kim, K.N.; Lee, W.-J. Heterostructured Co0.5Mn0.5Fe2O4-polyaniline nanofibers: highly efficient photocatalysis for photodegradation of methyl orange. *Korean J. Chem. Eng.* **2019**, *36*, 807–815. [CrossRef]
4. Sehar, S.; Naz, I.; Perveen, I.; Ahmed, S. Superior dye degradation using SnO2-ZnO hybrid heterostructure catalysts. *Korean J. Chem. Eng.* **2019**, *36*, 56–62. [CrossRef]
5. Trinh, T.K.; Truong, N.T.N.; Pham, V.T.H.; Kim, H.; Park, C. Effect of sulfur annealing on the morphological, structural, optical and electrical properties of iron pyrite thin films formed from FeS2 nano-powder. *Korean J. Chem. Eng.* **2018**, *35*, 1525–1531.
6. Chen, X.; Hou, Y.; Zhang, B.; Yang, X.H.; Yang, H.G. Low-cost SnS(x) counter electrodes for dye-sensitized solar cells. *Chem. Commun. (Camb).* **2013**, *49*, 5793–5795. [CrossRef]
7. Ettema, A.R.H.F.; de Groot, R.A.; Haas, C.; Turner, T.S. Electronic structure of SnS deduced from photoelectron spectra and band-structure calculations. *Phys. Rev. B* **1992**, *46*, 7363–7373. [CrossRef]
8. Minnam Reddy, V.R.; Gedi, S.; Park, C.; Robert, W.M.; Kotte, T.R. Development of sulphurized SnS thin film solar cells. *Curr. Appl. Phys.* **2015**, *15*, 588–598. [CrossRef]
9. Nguyen Tam Nguyen, T.; Hoang, H.H.T.; Trinh, T.K.; Pham, V.T.H.; Smith, R.P.; Chinho, P. Effect of post-synthesis annealing on properties of SnS nanospheres and its solar cell performance. *Korean J. Chem. Eng.* **2017**, *34*, 1208–1213.
10. Caballero, R.; Conde, V.; Leon, M. SnS thin films grown by sulfurization of evaporated Sn layers: Effect of sulfurization temperature and pressure. *Thin Solid Films* **2016**, *612*, 202–207. [CrossRef]
11. Lee, A.B.; Aron, W. Band alignment in SnS thin-film solar cells: Possible origin of the low conversion efficiency. *Appl. Phys. Lett.* **2013**, *102*, 132111.
12. Umme, F.; Khan, M.A.; Chinho, P. Elucidation of morphological and optoelectronic properties of highly crystalline chalcopyrite (CuInSe2) nanoparticles synthesized via hot injection route. *Korean J. Chem. Eng.* **2012**, *29*, 1453–1458.
13. Lee, H.; Jeong, D.S.; Mun, T.; Pejjai, B.; Minnam Reddy, V.R.; Anderson, T.J.; Park, C. Formation and characterization of CuInSe2 thin films from binary CuSe and In2Se3 nanocrystal-ink spray. *Korean J. Chem. Eng.* **2016**, *33*, 2486–2491. [CrossRef]
14. Lee, D.; Yong, K. Non-vacuum deposition of CIGS absorber films for low-cost thin film solar cells. *Korean J. Chem. Eng.* **2013**, *30*, 1347–1358. [CrossRef]
15. Lee, J.; Chen, H.; Koh, K.; Chang, C.L.; Kim, C.M.; Kim, S.H. Nanoassembly of CdTe nanowires and Au nanoparticles: pH dependence and reversibility of photoluminescence. *Korean J. Chem. Eng.* **2009**, *26*, 417–421. [CrossRef]
16. Greta, L.; Shun Li, S.; Neal R, K.; Tim, A.; Zi Kui, L. Thermodynamics of the S-Sn system: Implication for synthesis of earth abundant photovoltaic absorber materials. *Sol. Energy* **2016**, *125*, 314–323.

17. Ristov, M.; Sinadinovski, G.; Grozdanov, I.; Mitreski, M. Chemical deposition of Tin(II) sulphide thin films. *Thin Solid Films* **1989**, *173*, 53–58. [CrossRef]

18. Solar Frontier Achieves World Record Thin-Film Solar Cell Efficiency of 23.35%. Available online: http://www.solar-frontier.com/eng/news/2019/0117_press.html (accessed on 17 January 2019).

19. Elisa De, R. Cadmium telluride solar cells: Selenium diffusion unveiled. *Nat. Energy* **2016**, *1*, 16143–16146.

20. Rui, K.; Takeshi, Y.; Shunsuke, A.; Atsushi, H.; Kong Fai, T.; Takuya, K.; Hiroki, S. New world record Cu(In,Ga)(Se,S)2 thin film solar cell efficiency beyond 22%. In Proceedings of the 2016 IEEE 43rd Photovoltaic Specialists Conference (PVSC), Portland, OR, USA, 5–10 June 2016; pp. 1287–1291.

21. Mohamed, S.M.; Ibrahim, K.; Hmood, A.; Naser, M.A.; Falah, I.M.; Shrook, A.A. High performance near infrared photodetector based on cubic crystal structure SnS thin film on a glass substrate. *Mater. Lett.* **2017**, *200*, 10–13.

22. Ran, F.Y.; Xiao, Z.; Hiramatsu, H.; Ide, K.; Hosono, H.; Kamiya, T. SnS thin films prepared by H 2 S-free process and its p -type thin film transistor. *AIP Adv.* **2016**, *6*, 015112. [CrossRef]

23. Yao, K.; Li, J.; Shan, S.; Jia, Q. One-step synthesis of urchinlike SnS/SnS2heterostructures with superior visible-light photocatalytic performance. *Catal. Commun.* **2017**, *101*, 51–56. [CrossRef]

24. Huang, P.C.; Shen, Y.M.; Brahma, S.; Shaikh, M.O.; Huang, J.L.; Wang, S.C. SnSx (x = 1, 2) nanocrystals as effective catalysts for photoelectrochemical water splitting. *Catalysts* **2017**, *7*, 252. [CrossRef]

25. Afsar, M.F.; Rafiq, M.A.; Tok, A.I.Y. Two-dimensional SnS nanoflakes: synthesis and application to acetone and alcohol sensors. *RSC Adv.* **2017**, *7*, 21556–21566. [CrossRef]

26. Ou, J.Z.; Ge, W.; Carey, B.; Daeneke, T.; Rotbart, A.; Shan, W.; Wang, Y.; Fu, Z.; Chrimes, A.F.; Wlodarski, W.; et al. Physisorption-based charge transfer in two-dimensional SnS2 for selective and reversible NO2 gas sensing. *ACS Nano* **2015**, *9*, 10313–10323. [CrossRef]

27. Chung, R.J.; Wang, A.N.; Peng, S.Y. An enzymatic glucose sensor composed of carbon-coated nano tin sulfide. *Nanomaterials* **2017**, *7*, 39. [CrossRef]

28. Mudusu, D.; Nandanapalli, K.R.; Dugasani, S.R.; Kang, J.W.; Park, S.H.; Tu, C.W. Growth of single-crystalline cubic structured tin(II) sulfide (SnS) nanowires by chemical vapor deposition. *RSC Adv.* **2017**, *7*, 41452–41459. [CrossRef]

29. Wang, Y.; Huang, L.; Wei, Z. Photoresponsive field-effect transistors based on multilayer SnS 2 nanosheets. *J. Semicond.* **2017**, *38*, 034001. [CrossRef]

30. Manukumar, K.N.; Nagaraju, G.; Kishore, B.; Madhu, C.; Munichandraiah, N. Ionic liquid-assisted hydrothermal synthesis of SnS nanoparticles: Electrode materials for lithium batteries, photoluminescence and photocatalytic activities. *J. Energy Chem.* **2018**, *27*, 806–812. [CrossRef]

31. Li, J.; Xue, C.; Wang, Y.; Jiang, G.; Liu, W.; Zhu, C. Cu2SnS3 solar cells fabricated by chemical bath deposition-annealing of SnS/Cu stacked layers. *Sol. Energy Mater. Sol. Cells* **2016**, *144*, 281–288. [CrossRef]

32. Zhou, B.; Li, S.; Li, W.; Li, J.; Zhang, X.; Lin, S.; Chen, Z.; Pei, Y. Thermoelectric properties of SnS with na-doping. *ACS Appl. Mater. Interfaces* **2017**, *9*, 34033–34041. [CrossRef] [PubMed]

33. Jayalakshmi, M.; Mohan Rao, M.; Choudary, B.M. Identifying nano SnS as a new electrode material for electrochemical capacitors in aqueous solutions. *Electrochem. Commun.* **2004**, *6*, 1119–1122. [CrossRef]

34. Chauhan, H.; Singh, M.K.; Hashmi, S.A.; Deka, S. Synthesis of surfactant-free SnS nanorods by a solvothermal route with better electrochemical properties towards supercapacitor applications. *RSC Adv.* **2015**, *5*, 17228–17235. [CrossRef]

35. Gedi, S.; Minnam Reddy, V.R.; Kotte, T.R.; Park, Y.; Kim, W.K. Effect of C4H6O6 concentration on the properties of SnS thin films for solar cell applications. *Appl. Surf. Sci.* **2019**, *465*, 802–815. [CrossRef]

36. Gedi, S.; Minnam Reddy, V.R.; Alhammadi, S.; Guddeti, P.R.; Kotte, T.R.; Park, C.; Kim, W.K. Influence of deposition temperature on the efficiency of SnS solar cells. *Sol. Energy* **2019**, *184*, 305–314. [CrossRef]

37. Huang, C.C.; Lin, Y.J.; Chuang, C.Y.; Liu, C.J.; Yang, Y.W. Conduction-type control of SnSx films prepared by the sol-gel method for different sulfur contents. *J. Alloys Compd.* **2013**, *553*, 208–211. [CrossRef]

38. Minnam Reddy, V.R.; Pejjai, B.; Kotte, T.R.; Robert, W.M. X-ray photoelectron spectroscopy and X-ray diffraction studies on tin sulfide films grown by sulfurization process. *J. Renew. Sustain. Energy* **2013**, *5*, 031613. [CrossRef]

39. Sugaki, A.; Kitakaze, A.; Kitazawa, H. Synthesized tin and tin-silver sulfide minerals: Synthetic sulfide minerals (XIII). *Sci. Reports* **1985**, *3*, 199–211.

40. Mosburg, S.; Ross, D.R.; Bethke, P.M.; Toul-min, P. X-ray powder data for herzenbergite, teallite and tin trisulfide. *US Geol. Surv. Prof. Pap. C* **1961**, *424*, 347.

41. Guenter, J.R.; Oswald, H. New polytype form of tin (IV) sulfide. *Nat. Sci.* **1968**, *55*, 171–177.

42. Warren, B.E. *X-ray Diffraction*; Courier Corporation: Chelmsford, MA, USA, 1990; ISBN 0486663175.

43. Tallapally, V.; Nakagawara, T.A.; Demchenko, D.O.; Özgür, Ü.; Arachchige, I.U. Ge_{1-x} Sn_x alloy quantum dots with composition-tunable energy gaps and near-infrared photoluminescence. *Nanoscale* **2018**, *10*, 20296–20305. [CrossRef]

44. Gedi, S.; Minnam Reddy, V.R.; Park, C.; Jeon, C.W.; Kotte, T.R. Comprehensive optical studies on SnS layers synthesized by chemical bath deposition. *Opt. Mater. (Amst).* **2015**, *42*, 468–475. [CrossRef]

45. Biswajit, G.; Rupanjali, B.; Pushan, B.; Subrata, D. Structural and optoelectronic properties of vacuum evaporated SnS thin films annealed in argon ambient. *Appl. Surf. Sci.* **2011**, *257*, 3670–3676.

46. Nandanapalli, K.R.; Kotte, T.R. Preparation and characterisation of sprayed tin sulphide films grown at different precursor concentrations. *Mater. Chem. Phys.* **2007**, *102*, 13–18.

Article

PL Study on the Effect of Cu on the Front Side Luminescence of CdTe/CdS Solar Cells

Halina Opyrchal *, Dongguo Chen, Zimeng Cheng and Ken Chin

CNBM New Energy Materials Research Center, Department of Physics, New Jersey Institute of Technology, Newark, NJ 07102, USA

* Correspondence: opyrchal@njit.edu; Tel.: +1973-642-4283; Fax: +1973-596-5794

Received: 29 January 2019; Accepted: 4 July 2019; Published: 11 July 2019

Abstract: The effect of Cu on highly efficient CdTe thin solid film cells with a glass/TCO/CdS/CdTe structure subjected to $CdCl_2$ treatment was investigated by low-temperature photoluminescence (PL). The PL of the CdS/CdTe junction in samples without Cu deposition revealed a large shift in the bound exciton position due to the formation of CdS_xTe_{1-x} alloys with E_g (alloy) $\cong$ 1.557 eV at the interface region. After Cu deposition on the CdTe layer and subsequent heat treatment, a neutral acceptor-bound exciton (A^0_{Cu},X) line at 1.59 eV and two additional band-edge peaks at 1.54 and 1.56 eV were observed, indicating an increase in the energy gap value in the vicinity of the CdTe/CdS interface to that characteristic of bulk CdTe. These results may suggest the disappearance of the intermixing phase at the CdTe/CdS interface due to the presence of Cu atoms in the junction area and the interaction of the Cu with sulfur atoms. Furthermore, an increase in the intensity of CdS-related peaks in Cu-doped samples was observed, implying that Cu atoms were incorporated into CdS after heat treatment.

Keywords: photovoltaic; CdTe; CdS; solar cell; luminescence; spectroscopy

---

1. Introduction

The low-temperature photoluminescence (PL) of CdTe single crystals has been intensively studied in the past [1,2]. It is very well known that the typical low-temperature spectrum of undoped single-crystal CdTe consists of bands assigned to the recombination of free excitons (FX), bound excitons (AX or DX), free-to-bound transitions (FB), and donor–acceptor pair (DAP) transitions. Fewer low-temperature PL studies have been conducted for polycrystalline CdTe, which has been established as a very promising photovoltaic material for thin film solar cells [3,4]. A typical CdTe solar cell consisting of a CdS/CdTe layer grown on a SnO_2/glass substrate is based on recombination within an n-CdS/p-CdTe heterojunction. Despite a 9.7% lattice mismatch occurring between the hexagonal CdS and the cubic CdTe, high-efficiency devices result from this junction, probably due to interdiffusion at the CdS/CdTe interface [5–8] The S may passivate the grain boundaries and relieve the lattice mismatch between CdS and CdTe, thereby reducing recombination and improving device performance.

It is very well known that the post-deposition $CdCl_2$ annealing step is required to make reasonably efficient CdTe solar cells. It has been suggested that the $CdCl_2$ treatment enlarges grains, causes recrystallization, and enhances the diffusion of S from CdS into CdTe at the interface [2,7]. Additionally, the presence of chlorine dopants creates new recombination centers that may affect the recombination processes of photogenerated electron–hole pairs. It has been reported that the $CdCl_2$ treatment introduces two very important chlorine related defects, such as chlorine substituting on a tellurium site (Cl_{Te}) donor center and a V_{Cd}-Cl_{Te} (A) acceptor center that can be identified in the PL spectrum [1,5,9,10].

It was found recently that further improvement of CdTe solar cell can be achieved by Cu doping [11–16]. Cu enters the CdTe interstitially forming a shallow donor, Cu_i centers and gives rise to deep acceptor Cu_{Cd} centers via substitution [12–15].

In this paper, results of studies on the effect of Cu doping on the low-temperature front side PL of CdS/CdTe solar cells will be presented.

2. Experimental Design

2.1. Film Deposition and Cell Fabrication

Cadmium telluride solar cells were fabricated in the superstrate configuration with the structure: glass/TCO/n$^+$-CdS/p-CdTe/graphite/metal back contact in our laboratory (CNBM Center). Commercially available soda-lime glass coated with SnO$_2$:F/HRT was used as a substrate for the depositions. CdS (~80 nm in thickness) was deposited by chemical bath deposition (CBD) at 88 °C using cadmium chloride, thiourea, ammonium acetate, and ammonia, followed by annealing at 400 °C. CdTe (6–10 µm in thickness) was deposited by close-spaced sublimation (CSS) at T_s = 600 °C using graphite susceptors in 10–15 Torr He/O$_2$. The CdTe was then soaked in CdCl$_2$/methanol at 80 °C, followed by a furnace anneal under controlled conditions (380 °C, He/O$_2$, 300 Torr) in order to improve the CdTe structure and minority carrier lifetime. To form the back contact, a nitric-phosphoric (NP) acid etch was used to remove the surface oxide, and a very thin layer of Cu was evaporated. The efficiency of the cells examined in this work was in the range of 10%–11%. The overall process is described in [17].

A schematic of a typical CdS/CdTe solar cell is shown in Figure 1.

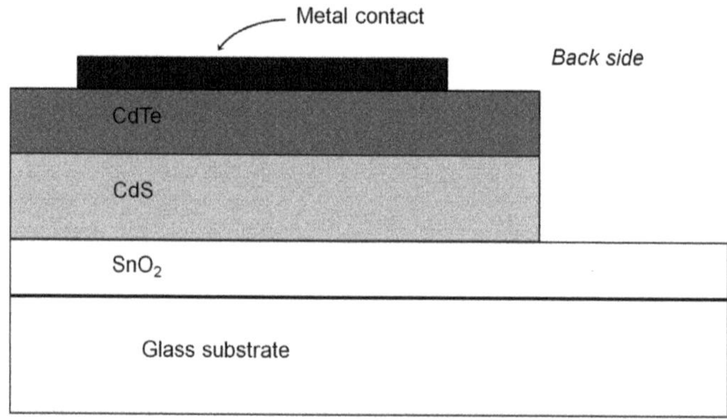

Figure 1. The glass substrate represents the front side of a solar cell and the CdTe layer represents the backside of a solar cell. The glass substrate, SnO$_2$, and CdS layers are transparent to a 658 nm line that is absorbed only by the CdTe film at the interface (junction luminescence).

2.2. Photoluminescence Measurements

The PL spectra were measured for undoped and Cu-doped CdTe/CdS cells produced from the CdTe source with a purity of 99.999% (5N) or 99.99999% (7N) and illuminated by monochromatic light directed onto the backside or front side of the solar cell structure. All measurements were performed on CdCl$_2$-treated cell structures, but before the deposition of the back contact.

For the backside PL, the measurements were taken by illuminating the CdTe/CdS cells with a 658 nm line generated by a red laser diode (LPM658, Newport, Irvine, CA, USA). For the front side luminescence, in addition to red excitation, a blue line with a wavelength of 405 nm (laser diode LQA405-8SE, Newport, Irvine, CA, USA) was also used. In Cu-doped samples, Cu was introduced to the cell through diffusion at 400 °C from a layer of pure metal deposited on the CdTe after chlorine treatment. The thickness of the copper layer was 10 nm. All of the samples used for PL measurements were placed in a closed-cycle cryostat with a controlled temperature ranging from 10 to 300 K. Spectra were recorded using a grated monochromator coupled to a cooled photomultiplier tube (R106999, Hamamatsu, Tokyo, Japan).

3. Experimental Results and Discussion

3.1. Optical Characterization of the CdS/CdTe Solar Cells

In Figure 2, the results of the backside PL excited by 658 nm light are shown for two different CdS/CdTe samples. One sample was a solar cell grown in the NREL laboratory (Golden, Colorado, USA) and kindly supplied to us for testing. The second sample was grown in the CNBM Center using the same methodology used to generate solar cell devices with an efficiency of about 11%. As can be seen in Figure 2, the spectra obtained from both samples were similar. The presence of an excitonic band with a maximum at 1.59 eV indicated good-quality CdTe polycrystalline films in both samples. In both samples, defect-related peaks observed at around 1.44 eV were recorded. The defect-related peaks were assigned to DAP transitions between the Cl_{Te} donor centers and the A acceptor centers [1,10]. The origin of the peaks between 1.55 and 1.50 eV in both samples is still under discussion. A similar emission behavior was reported by other authors [1,10] who assigned bands in this region to DAP transitions (1.56 eV) and free electron–acceptor (e, A^0) transitions (1.54 eV).

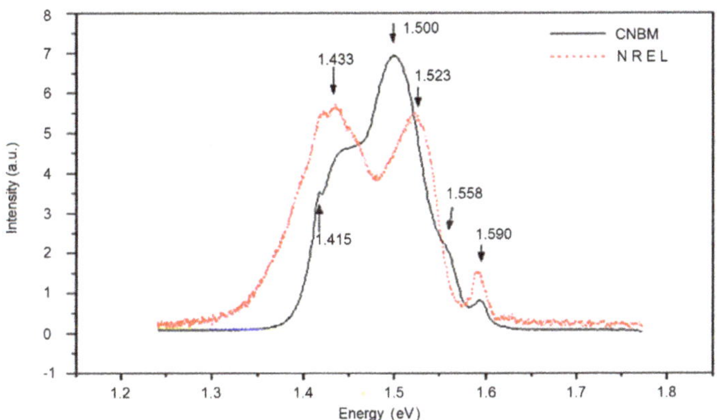

Figure 2. Backside luminescence (10 K) of CdS/CdTe solar cells grown in the NREL laboratory and CNBM New Energy Materials Research Center, respectively.

3.2. Front Side Luminescence of CdS/CdTe Structure Excited by a 658 nm (1.88 eV) Line

For front side luminescence, solar cells were illuminated from the front side of the device consisting of glass covered by TCO and a CdS window transparent to excitation light of 658 nm. However, the excitation light was strongly absorbed by the CdTe layer. Due to the high absorption, the generation of free electron–hole pairs and their subsequent separation or recombination occurred exclusively in a very narrow layer of the CdTe adjacent to the CdS/CdTe interface. Front side PL, called junction luminescence, gives us direct information about processes occurring in this area.

3.2.1. Front Side Luminescence of an Undoped CdS/CdTe

Figure 3 shows the PL spectrum obtained for the undoped CdS/CdTe structure excited from the junction side through the glass. The backside PL spectrum obtained for the same sample is shown for comparison. The junction PL of undoped CdS/CdTe structures has been reported by several investigators [10,11,18,19] and results similar to ours were obtained. The front side spectrum consists of two peaks at 1.53 and 1.43 eV; the high-energy peak at 1.59 eV detected in the backside spectrum is not present. There are two possible explanations for this observation. One possibility is to assume that the origin of the 1.53 eV peak in the junction PL and the 1.59 eV peak in the backside PL is the same, namely that they are both due to bound-exciton emission. The 73 MeV shift in the exciton peak towards a lower energy for the junction region can be attributed to the reduction of the bandgap due to the formation of the CdS_xTe_{1-x} alloy at the interface described in [7–10]. The shift found in this paper is similar to that found in [19,20].

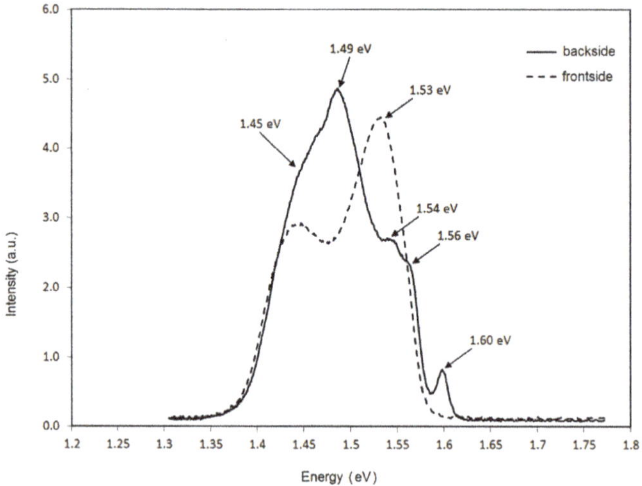

Figure 3. Photoluminescence (PL) spectra (10 K) from undoped CdS/CdTe solar cells excited from the backside (full line) and front side (dotted line) of the cell.

The alternative explanation is to assume that the excitonic band at 1.59 eV is depressed due to a high density in defects at the interface and that the 1.53 eV peak is equivalent to the broad free-to-bound peak around 1.56 eV detected in the same sample excited from the back. However, the analysis of the temperature dependence of the intensity of the 1.53 eV peak presented in the next section strongly favors the assertion of an excitonic origin of this peak.

3.2.2. Temperature Dependence of the 1.53 eV Peak

In order to further clarify the origin of the 1.53 eV peak found in the junction PL of the undoped sample, additional measurements of PL as a function of temperature were conducted. The analysis of the temperature dependence of the intensity of the 1.53 eV transition presented in Figure 4 shows a good fit with a two activation-energy model:

$$\frac{I(T)}{I(0)} = \frac{1}{1 + C_1 e^{\frac{-E_1}{kT}} + C_2 e^{\frac{-E_2}{kT}}} \tag{1}$$

proposed by Bimberg et al. [20] for the temperature decay of bound excitons. The activation energies E_1 and E_2 in this equation represent the electron–hole binding energy and the binding energy of the exciton to the impurity, respectively.

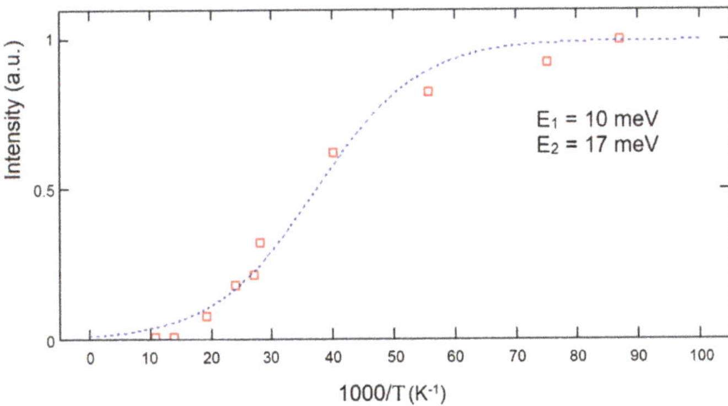

Figure 4. Temperature dependence of the PL intensity of the 1.53 transition in undoped cells along with the two-activation energy fit.

The binding energy of 10 meV is in good agreement with the binding energy of the free exciton [1], and it is reasonable to assume that the 17 meV point represents the binding energy of the exciton to the unidentified acceptor or the chlorine donor. The exciton position of 1.53 eV and the two binding energies, 0.01 and 0.017 eV, can be used to calculate the energy gap for CdS_xTe_{1-x} alloy [1]:

$$E_g \text{ (alloy)} = 1.53 + 0.01 + 0.017 = 1.557 \text{ eV} \tag{2}$$

Following the work of Ohata [21] and Pal [22], the composition of such alloy can be predicted using a simple quadratic equation:

$$E_{g\ (CdSxTe1-x)} = kx^2 + (E_{g\ (CdS)} - E_{g\ (CdTe)} - k)x + E_{g\ (CdTe)} \tag{3}$$

Assuming $k = 1.7$ eV from [21] and substituting a 10 K bandgap for CdS ($E_{g\ (CdS)} = 2.59$ e) and for that of CdTe ($E_{g\ (CdTe)} = 1.61$ eV), the value $x \cong 0.08$ can be obtained.

3.2.3. Effect of Cu on the Front Side Luminescence of CdS/CdTe Solar Cells

The effect of Cu on the solar cell structure was studied by measuring the front side PL from a sample with a 10 nm Cu layer deposited on the CdTe side of the cell that was subjected to chlorine treatment, as described previously. After Cu deposition, the cell was annealed for 10 min at 400 °C. As can be seen from the Figure 5, in addition to peaks at 1.43 and 1.53 eV observed in the undoped samples, two new peaks at 1.595 and 1.56 eV were recorded in the Cu-doped samples. The following conclusions can be drawn from the comparison of the spectra taken for doped and undoped samples:

- The PL results show that Cu was present in the vicinity of the junction as a result of its migration from the backside of the solar cell during the post deposition annealing.
- The presence of peaks at maximums of 1.59 and 1.56 eV (absent in copper-free cells) indicated that the composition of the CdTeS alloy at the interface was dramatically changed due to Cu diffusion into the junction area. It has already been established [23,24] that the peak at 1.59 eV in Cu-doped CdTe corresponds to Cu_{Cd}-bound exciton.
- The position of the exciton peak can be used to estimate the value of the energy gap in the CdTe layer adjacent to the CdTe/CdS interface using the following relation: $h\omega$ bound exciton $= E_g - E_B$. The value of E_g obtained from the junction luminescence of Cu-doped sample is essentially identical to the $E_g = 1.605$ of bulk CdTe in 10 K [25]. This indicates that the formation of the CdSTe alloy observed in undoped solar cells is suppressed by the presence of copper in the junction area.

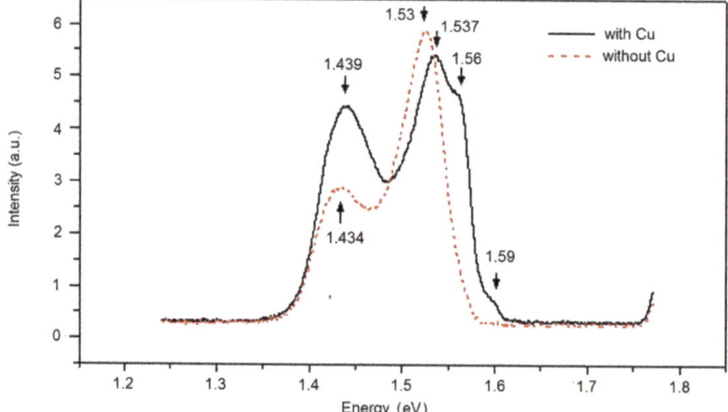

Figure 5. Front side luminescence of Cu-doped (full line) and undoped (dotted line) CdS/CdTe solar cells taken at 10 K.

It is worth noting that a similar effect has not been found in previously reported studies [11,18,19] of front side luminescence of Cu-doped CdTe/CdS solar cells. This may suggest that the Cu did not reach the junction area in those samples.

3.3. Front Side Luminescence of CdS/CdTe Structure Excited by a 405 nm (3.06 eV) Line

3.3.1. Front Side Luminescence of an Undoped CdS/CdTe Structure

Contrary to red light, blue light is strongly absorbed by the CdS layer. Therefore, it is reasonable to assume that the PL shown in Figure 6 is related mainly to emission from the CdS layer. In a high energy range, four peaks with maximal emission at 2.34, 1.92, 1.79 eV and a weak emission at 1.65 eV were recorded. Despite the extensive literature on the luminescence of polycrystalline thin film CdS, the interpretation of the observed PL bands may not be easy. The low-temperature PL strongly depends on the type of deposition and the post-deposition treatment [26,27]. According to the literature data [26], it is reasonable to assume that the emission at around 2.34 eV can be attributed to the excitons bound to substitutional Te atoms on the sulfur lattice site while the two bands at 1.79 and 1.65 eV may be due to a DAP transition involving Te complexes in the CdS. The peak at 1.92 eV had a similar position to what has been called a red band, ascribed to V_{Cd}-V_s defects [26].

In the low energy range, two strong peaks at 1.51 and 1.41 eV were detected. The origin of those peaks is not clear. They may be related to an emission from CdTe as the position of those peaks are similar to that recorded for the front side PL excited by red light. On the other hand, similar peaks were found in polycrystalline CdS grown on glass or TCO by chemical bath deposition [26], and were ascribed to the transition of electrons trapped in surface states to the valence band.

3.3.2. Effect of Cu on the Front Side Luminescence of CdS/CdTe Solar Cells

The effect of Cu doping on CdS luminescence has never been studied before. As can be seen from Figure 6, the intensity of all four high energy peaks increased significantly due to Cu doping but no peak shift was observed. This can be understood assuming that the Cu passes through the interface and diffuses into the CdS film, where it reacts with native CdS defects in such way that the number of nonradiative recombination events decreases. The decrease in the intensity of the peaks at 1.51 and 1.41 eV in doped samples suggests that those peaks are related to the CdS lattice rather than to the CdTe. The reduction of the intensity of the peaks can be ascribed to the reduction in the number of surface defects in the Cu-doped sample.

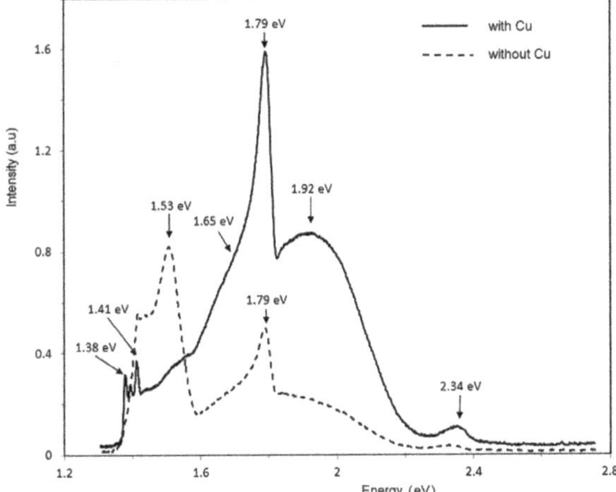

Figure 6. Front side luminescence (10 K) from Cu-doped (solid line) and undoped (dotted line) CdS/CdTe solar cells excited by the 405 nm line.

4. Conclusions

Low-temperature junction PL has been used to study the effect of Cu deposited on CdTe on the electronic structure of CdTe/CdS solar cells grown in our laboratory.

In undoped cells, good evidence for the interdiffusion of sulfur and bandgap reduction was found in CdTe/CdS solar cells subjected to $CdCl_2$ treatment. The energy gap of 1.557 eV was calculated and the composition of the interface alloy was estimated to be $x \cong 0.08$.

In Cu-doped samples, the junction PL data provides evidence that Cu diffuses into the junction area, forming Cu_{Cd} centers. The most interesting result of this work is the fact that the energy bandgap of CdTe at the interface restores its original value of about 1.60 eV. This allows us to suggest that the presence of Cu in the junction area hinders the interdiffusion of sulfur through the interface. If so, the effect of sulfur diffusion on the cell performance should be reexamined.

Finally, this is the first time that CU has been documented to diffuse through the interface into the CdS layer, changing the character of the recombination processes in the CdS.

Author Contributions: Conceptualization, H.O. and D.C.; Methodology, H.O.; Validation, H.O.; Formal Analysis, H.O. and D.C.; Investigation, D.C.; resources, Z.C.; Data Curation, H.O.; Writing—Original Draft Preparation, H.O.; Writing—Review and Editing, H.O., D.C. and Z.C.; Visualization, D.C. and H.O.; Supervision, K.C.; Project Administration, H.O. and K.C.; Funding Acquisition, K.C.

Funding: This research was funded by China Triumph International Engineering CO. LTD.

Acknowledgments: The authors thank the National Renewable Energy Laboratory, Golden, Colorado 80401 for supplying CdTc/CdS solar cells for reference measurements and members of the NJIT CNBM research center for processing help and useful discussions. In particular, the authors wish to thank Zhitau Wang for his involvement in taking PL measurements for this work.

Conflicts of Interest: The authors declare no conflict of interest.

References

1. Liu, X. Photoluminecence and Extended X-ray Absoprtion Fine Structure Studies on CdTe Material. Ph.D. Thesis, University of Toledo, Toledo, OH, USA, 2006.
2. Dhere, R.G. Study of the CdS/CdTe Interface and Its Relevance to Solar Cell Properties. Ph.D. Thesis, University of Colorado, Boulder, CO, USA, 1997.

3. Wu, X. High-efficiency polycrystalline CdTe thin film solar cells. *Sol. Energy* **2004**, *77*, 803–814. [CrossRef]

4. Bosio, A.; Rosa, G.; Romeo, N. Past, present, and future of the thin film CdTe/CdS solar cells. *Sol. Energy* **2018**, *175*, 31–43. [CrossRef]

5. Potter, M.D.G.; Cousins, M.; Durose, K.; Halliday, D.P. Effect of interdiffusion and impurities on thin film CdTe/CdS photovoltaic junction. *J. Mater. Sci. Mater. Electron.* **2000**, *11*, 525–530. [CrossRef]

6. Moutinho, H.R.; Dhere, R.G.; Jiang, C.S.; Albin, D.S.; Al-Jassim, M.M. Electrical properties of DcTe/CdS and CdTe/SnO$_2$ solar cells studied with scanning kelvin probe microscopy. In Proceedings of the 35th Photovoltaic Specialists Conference, Honolulu, HI, USA, 20–25 June 2010; p. 1955.

7. Metzger, W.K.; Albin, D.; Romero, M.J.; Dippo, P.; Young, M. CdCl$_2$ treatment, S diffusion, and recombination in polycrystalline CdTe. *J. Appl. Phys.* **2006**, *99*, 103703–103708. [CrossRef]

8. Delahoy, A.E.; Cheng, Z.; Chin, K. Evidence for CdTe$_{1-x}$S$_x$ Compound formation in CdTe solar cells from high-precision, temperature-dependent device measurements. In Proceedings of the 39nd Photovoltaic Specialist Coonference, Tampa, FL, USA, 16–21 June 2013; pp. 1945–1948.

9. Van Gheluwe, J.; Versluys, D.; Poelman, P.; Clauws, P. Photoluminescence study of polycrystalline CdS/CdTe thin solar cells. *Thin Solid Film.* **2005**, *480*, 264–268. [CrossRef]

10. Okamoto, T.; Matsuzaki, Y.; Amin, N.; Yamada, A.; Konagai, M. Characterization of highly efficient CdTe thin film solar cells by low-temperature photoluminescence. *Jpn. J. Appl. Phys.* **1998**, *7*, 3894–3899. [CrossRef]

11. Okamoto, T.; Yamada, A.; Konagai, M. Optical and electrical characterization of highly efficient CdTe thin film solar cells. *Thin Solid Film.* **2001**, *387*, 6–10. [CrossRef]

12. Grecu, D.; Compan, A.D.; Young, D.; Jayamaha, U.; Rose, D.H. Photoluminescence of Cu-doped CdTe and related stability issues in CdS/CdTe solar cells. *J. Appl. Phys.* **2000**, *88*, 2490–2496. [CrossRef]

13. Demtsu, S.H.; Albin, D.S.; Sites, J.R.; Metzger, W.K.; Duda, A. Cu-related recombination in CdS/CdTe solar cell. *Thin Solid Film.* **2008**, *516*, 2251–2254. [CrossRef]

14. Okamoto, T.; Hayashi, R.; Hara, S.; Ogawa, Y. Cu doping of CdTe layers in polycrystalline CdTe thin-film solar cells for top cells of multijunction solar cells. *Jpn. J. Appl. Phys.* **2013**, *52*, 102301. [CrossRef]

15. Perrenoud, J.; Kranz, L.; Gretener, C.; Pianezzi, F.; Nishiwaki, A.; Buecheler, S.; Tiwari, A.N. A comprehensive picture of cu doping in CdTe solar cells. *J. Appl. Phys.* **2013**, *114*, 174505–174525. [CrossRef]

16. Kuciauskas, D.; Kanevce, A.; Duenow, J.N.; Dippo, P.; Young, M.; Li, J.V.; Levi, D.H.; Gessert, T.A. Spectrally and time resolved photoluminescence analysis of the CdS/CdTe interface in thin-film photovoltaic solar cells. *Appl. Phys. Lett.* **2013**, *102*, 173902. [CrossRef]

17. Kharangarh, P.R.; Misra, D.; Georgiou, G.E.; Chin, K.K. Characterization of space charge layer deep defects in n$^+$-CdS/p-CdTe solar cells by temperature dependent capacitance spectroscopy. *J. Appl. Phys.* **2013**, *113*, 144504–144509. [CrossRef]

18. Kuciauskas, D.; Dippo, D.; Kanevce, A.; Zhao, Z.; Cheng, L.; Los, A.; Gloeckler, M.; Metzger, W.K. The impact of Cu on recombination in high voltage CdTe solar cells. *Appl. Phys. Lett.* **2015**, *107*, 243906. [CrossRef]

19. Kuciauskas, D.; Dippo, D.; Zhao, Z.; Cheng, L.; Kanevce, A.; Metzger, W.K.; Gloeckler, M. Recombination analysis in cadmium telluride photovoltaic solar cells with photoluminecence spectroscopy. *IEEE J. Photovolt.* **2016**, *6*, 313–319. [CrossRef]

20. Bimberg, D.; Sondergeld, M.; Grobe, E. Thermal dissociation of excitons bounds to neutral acceptors in high-purity GaAs. *Phys. Rev. B* **1971**, *4*, 3451–3455. [CrossRef]

21. Ohata, K.; Saraie, J.; Tanaka, T. Optical energy gap of the mixed crystal CdS$_x$Te$_{1-x}$. *Jpn. J. Appl. Phys.* **1973**, *12*, 1641. [CrossRef]

22. Pal, R.; Dutta, J.; Chaudhuri, S.; Pal, A. CdS$_x$Te$_{1-x}$ films: Preparation and properties. *J. Phys. D Appl. Phys.* **1993**, *26*, 704–710. [CrossRef]

23. Molva, E.; Chamonal, J.P.; Saminadayar, K.; Pajot, B.; Neu, G. Excited states of Ag an Cu acceptors in CdTe. *Solid State Commun.* **1982**, *44*, 351–355. [CrossRef]

24. Laurenti, J.P.; Bastide, G.; Rouzeyre, M.; Triboulet, R. Localized defects in p-CdTe: Cu doped by copper incorporation during Bridgman growth. *Solid State Commun.* **1988**, *67*, 1127–1130. [CrossRef]

25. Fonthal, G.; Tirado-Mejia, L.; Marin-Hurtado, J.I.; Ariza-Calderon, H.; Mendoza-Alvarez, J.G. Temperature dependence of the band gap energy of crystalline CdTe. *J. Phys. Chem. Solids* **2000**, *61*, 579–583. [CrossRef]

26. Hernandez-Fenellosa, M.A.; Halliday, D.P.; Durose, K.; Campo, M.D.; Beier, J. Photoluminescence studies of CdS/CdTe solar cells treated with oxygen. *Thin Solid Film.* **2003**, *431–432*, 176–180. [CrossRef]
27. Abken, A.; Halliday, D.P.; Durose, K. Photoluminescence study of polycrystalline CdS thin film layers grown by closed-spaced sublimation and chemical bath deposition. *J. Appl. Phys.* **2009**, *105*, 064515–064524. [CrossRef]

 *coatings*

Article

Factors Affecting Electroplated Semiconductor Material Properties: The Case Study of Deposition Temperature on Cadmium Telluride

A.A. Ojo [1,2,*] and I. M. Dharmadasa [2]

[1] Department of Mechanical Engineering, Ekiti State University (EKSU), Ado-Ekiti 360211, Nigeria
[2] Electronic Materials and Sensors Group, Materials and Engineering Research Institute (MERI), Sheffield Hallam University, Sheffield S1 1WB, UK; dharme@shu.ac.uk
* Correspondence: chartell2006@yahoo.com or ayotunde.ojo@eksu.edu.ng

Received: 9 April 2019; Accepted: 4 June 2019; Published: 7 June 2019

Abstract: Electrodeposition of cadmium telluride (CdTe) on fluorine doped tin oxide (FTO) using two electrode configurations was successfully achieved with the main focus on the growth temperature. The electroplating temperatures explored ranged between 55 and 85 °C for aqueous electrolytes containing 1.5 M cadmium nitrate tetrahydrate ($Cd(NO_3)_2 \cdot 4H_2O$) and 0.002 M tellurium oxide (TeO_2). The ensuing CdTe thin-films were characterized using X-ray diffraction (XRD), UV-Vis spectrophotometry, scanning electron microscopy (SEM), energy dispersive X-ray (EDX), and photoelectrochemical (PEC) cell measurements. The electroplated CdTe thin-films exhibit a dominant (111) CdTe cubic structure, while the crystallite size increases with the increase in the electroplating temperature. The dislocation density and the number of crystallites per unit area decrease with increasing growth temperature. The optical characterization depicts that the CdTe samples show comparable absorbance and a resulting bandgap of 1.51 ± 0.03 eV for as-deposited CdTe layers. A marginal increase in the bandgap and reduction in the absorption edge slope towards lower deposition temperatures were also revealed. The annealed CdTe thin-films showed improvement in the energy bandgap as it tends towards 1.45 eV while retaining the aforementioned absorption edge slope trend. Scanning electron microscopy shows that the underlying FTO layers are well covered with increasing grain size observable relative to the increase in the deposition temperature. The energy dispersive X-ray analyses show an alteration in the Te/Cd relative to the deposition temperature. Higher Te ratio with respect to Cd was revealed at deposition temperature lower than 85 °C. The photoelectrochemical cell study shows that both *p*- and *n*-type CdTe can be electroplated and that deposition temperatures below 85 °C at 1400 mV results in *p*-type CdTe layers.

Keywords: electrodeposition; CdTe film; two-electrode configuration; thin films; electroplating temperature

---

1. Introduction

Electrodeposition has emerged as a one of the versatile and cost-effective growth techniques of metal, metalloid, and semiconductor materials [1]. With emphasis on semiconductor growth, the use of either two- or three- electrode electroplating technique has been effective in the growth of high-performance semiconductor materials [2]. Aside from the the electroplating configuration, challenging factors affecting the reproducibility of electroplated materials include the solutes incorporated and solvent utilized, electrolytic bath ionic concentrations, solution aging (electroplating age), the stabilities of both growth pH and deposition temperature, and the deposition current density [3]. Provided these factors are optimized, high-quality semiconductor materials such as cadmium telluride (CdTe) thin-films can be grown and incorporated for different applications. CdTe is one of the II-VI semiconductor materials that has been grown using several techniques including electroplating [4,5] and has been extensively researched owing to its properties [6]. Due to the direct bandgap of CdTe at room temperature (~1.45 eV), it is capable of absorbing a substantial fraction of the electromagnetic spectrum under AM1.5 conditions. This characteristic has been explored in the photovoltaic (PV) community in achieving high-efficiency CdTe-based solar cells [7]. The recent hike in the conversion efficiency from 16.5% in 2004 [8] to 22.1% in 2016 [9] as reported in the literature is mainly due to the eradication of defects within the crystal lattice or traps within the bandgap in addition to improved crystallinity, passivation of the grain boundaries which is due to better understanding of both material and device issues. Therefore, it is fundamental to strive towards process optimization amongst others. Under both two and three- electrode electrodeposition configurations, different deposition temperatures have been utilized by independent researchers [10–16] without a unifying examination of the effect of electroplating temperature in aqueous solution. Although, limiting factors such as the operating temperature of reference electrode (of about 70 °C) is to be considered for the three-electrode configurations. Therefore, this publication examines the effect of temperature of the electrolyte on the structural, optical, morphological compositional properties and electronic properties of electrodeposited CdTe layers grown from an aqueous solution containing tellurium dioxide (TeO$_2$) and cadmium nitrate Cd(NO$_3$)$_2$ as the respective precursors of Te and Cd.

2. Materials and Methods

2.1. Thin-Film Synthesis

Cadmium telluride thin-films were electrodeposited from a solution containing 1.5 M cadmium nitrate tetrahydrate (Cd(NO$_3$)$_2$·4H$_2$O) and 0.002 M tellurium oxide (TeO$_2$). The respective precursors for Cd and Te were dissolved in 400 mL of deionized (DI) water contained in a polypropylene beaker using the set up depicted in Figure 1. The resulting aqueous solution will be referred to as CdTe-bath henceforth in this report. Using these precursor concentrations and electrolytic cell set up, four electrolytic baths were created. To achieve a transparent and homogenous solution, the baths were stirred for about 300 min.

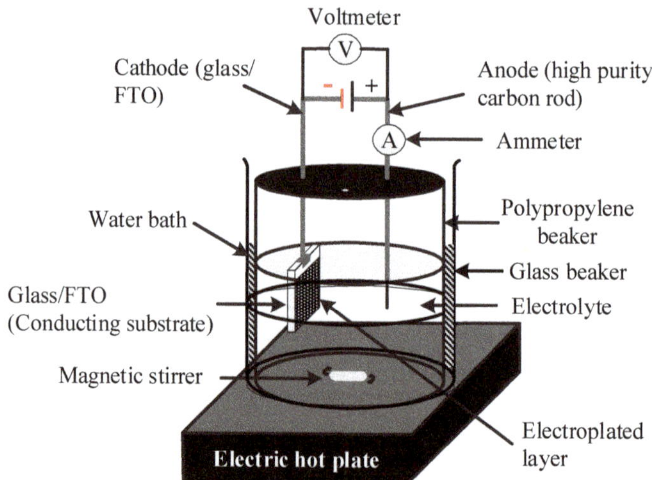

Figure 1. Typical two-electrode electrodeposition configuration.

Prior to electroplating, the pH and the magnetic stirring rate of the bath were maintained at 2.00 ± 0.02 and ~300 rpm. The adjustment of the pH is achieved by either using nitric (HNO$_3$) acid to lower the pH value or ammonium hydroxide (NH$_4$OH)—an alkaline to increase the pH value.

The deposition temperature is set at 55, 65, 75, and 85 °C for respective electrolytic baths, while the deposition voltage is kept constant at 1400 mV for all the CdTe layers. The 1400 mV cathodic growth voltage is based on prior optimization of CdTe thin-films grown from a pure $Cd(NO_3)_2$ electrolyte as described in the literature [17]. The deposition temperature lower or higher than this range were not reported due to low adhesion of the ED-CdTe thin-film and the formation of water bubbles on the g/FTO substrate due to the close proximity of the deposition temperature to the boiling point of water.

The glass/fluorine-doped tin oxide (g/FTO) substrates utilized as the working electrode were cut into strips with a dimension of 3×2 cm^2. The g/FTO strips were washed using soap water in an ultrasonic bath, alcohol, rinse thoroughly in DI water and dried using nitrogen gas [18]. After the thin-film layer deposition, the 3×2 cm^2 strips were cut into two halves of 3×1 cm^2. One half is left as-deposited (AD) while the other is cadmium chloride treated (CCT) to improve the material and electronic properties of the CdTe layers as it used in the CdTe-based photovoltaic device-ready process. The CCT treatment was performed by adding a few drops of aqueous solution containing 0.1 M $CdCl_2$ in 20 mL of DI water to the CdTe surface. The full coverage of the CdTe layers with the $CdCl_2$ solution was achieved by spreading the solution using solution-damped cotton bud. The CdTe layers were allowed to air-dry and annealed afterwards in air at 400 °C for 20 min. The CCT-CdTe layers were allowed to cool in the air before both the AD and the CCT layers were rinsed to eliminate the loosely adhered particles of CdTe, Cd, and Te from the surfaces of the thin-film and dried in the presence of nitrogen gas.

It is necessary to note that cyclic voltammetry as shown in Figure 2 of the CdTe-bath was performed at electrolytic bath conditions when bath temperature, pH, and stirring rate were set at 85 °C, 2.00 ± 0.02 and ~300 rpm. The voltammogram is the current-voltage (I-V) curve of the CdTe-bath to determine the optimal cathodic voltage region for the deposition of CdTe.

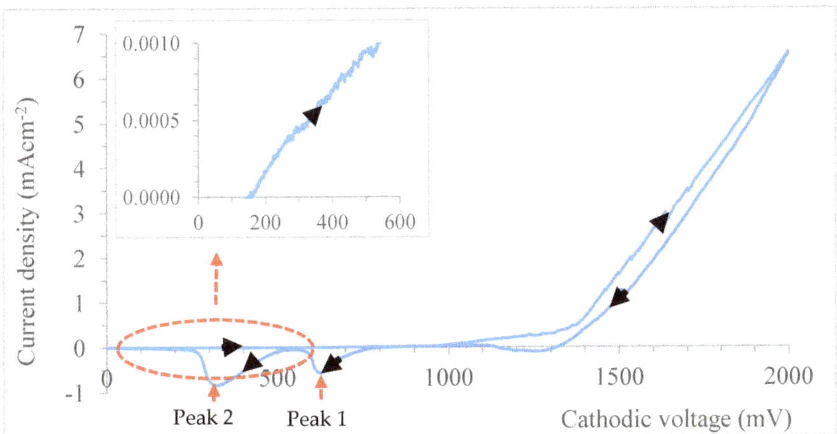

Figure 2. Typical cyclic voltammogram of an aqueous electrolyte at 85 °C containing 1.5 M $Cd(NO_3)_2 \cdot 4H_2O$ and 0.0002 M TeO_2 in 400 mL of DI water.

Tellurium is deposited first because it is more positive as compared to cadmium with respective standard reduction potential value of +593 mV and −403 mV with respect to standard H_2 electrode. From the inset of Figure 2, Te starts depositing at a cathodic voltage of ~170 mV and above, while Cd starts deposition at 1000 mV. Details of the optimization process have been documented by the author's group in the literature [12,17].

2.2. Thin-Film Characterization

The electroplated layer characterisation were performed using equipment made available by the Materials and Engineering Research Institute (MERI) research institute, Sheffield Hallam University (SHU), UK.

The structural characterization was carried out using Philips PW 3710 X'pert diffractometer (Almelo, Netherlands) mounted with 1.5406 Å wavelength Cu-Kα monochromator. The X-ray diffraction system is also equipped with X'Pert High Score which aided with phase identification, dominant diffraction and for the estimation of the crystallite sizes. For the experiments presented in this work, thin-films were scanned between the range of $2\theta = 20°$–$70°$. Using the Scherrer formula as shown in Equation (1), the crystallite size D was estimated. Where β denotes the full-width-at-half-maximum (FWHM) of the diffraction intensity in radians, θ denotes the Bragg angle, λ denotes the X-rays wavelength (which is 1.5406 Å for Cu-Kα monochromator) and K is the shape constant. For spherical geometry, K is taken as 0.94.

$$D = \frac{K\lambda}{\beta \cos \theta} \tag{1}$$

The dislocation density, δ, which defines the length of dislocation lines per unit volume of crystal in the thin-film is calculated using Equation (2) as reported in the literature [19]

$$\delta = \frac{1}{D^2} \tag{2}$$

The estimation of the number of crystallites per unit area in the thin-film, N, was done using Equation (3) as documented in [19,20].

$$N = \frac{t}{D^3} \tag{3}$$

The film thickness was measured using a UBM 1080 Microfocus optical profilometer (UBM Messtechnik, Koln, Germany) and mathematically estimated using Faraday's law of electrolysis (see Equation (4)), where T is the film thickness, J is the average current density during deposition, M is the molar mass of CdTe ($M_{\text{CdTe}} = 240.01$ gmol^{-1}), t is the duration of deposition, n is the number of electrons transferred for deposition of 1 molecule of CdTe ($n = 6$), and F is the constant defined by Faraday as 96,485 Cmol^{-1}, and ϱ is the density of CdTe.

$$T = \frac{JMt}{n\varrho F} \tag{4}$$

The optical absorbance data of the electroplated thin-films were taking within the range of 200 to 800 nm using Cary50 Scan UV-visible spectrophotometer (Agilent Technologies, Santa Clara, CA, USA). From the absorbance data accumulated within the specified wavelength range, the bandgap was estimated using the Tauc's formula illustrated in Equation (5) via the graph of $(\alpha h v)^2$ versus $(h v)$. Where the coefficient of absorption is represented by α, the bandgap energy is represented by E_g, the Planck's constant is represented by h, the incident photon frequency is represented by v, the proportionality constant which depends on the refractive index of the sample under investigation is represented by k, and m equals 0.5 for a direct bandgap semiconductor. The extrapolation of the straight line portion of the Tauc's plot (at $(\alpha h v)^2 = 0$) gives the bandgap energy.

$$\alpha = \frac{k\left(h v - E_g\right)^m}{h v} \tag{5}$$

The examination of the thin film's surface morphology and composition was performed in vacuum condition using FEI Nova200 NanoSEM (Electron Nanoscopy Instrument, Lincoln, NE, USA) fitted with energy dispersive X-ray (EDX) detector. The confirmation of the electrical conduction type is done using photoelectrochemical (PEC) cell measurements by the formation of a junction between the solid (g/FTO/CdTe) and liquid (an aqueous solution of 0.1 M $Na_2S_2O_3$ in 20 mL DI water). The comprehensive detail of the PEC set up is incorporated in [18].

3. Result and Discussion

Material Characterization: In order to validate the results in this Section, all the stated values correspond to the average of three replica samples investigated under the same conditions.

3.1. X-Ray Diffraction (XRD) Analysis

For this set of experiments, the thicknesses of the CdTe layers electrodeposited at a different temperature from CdTe-baths was maintained at ~1μm under as-deposited conditions using factors such as deposition time, stirring rate amongst others. Figure 3a,b illustrate the typical XRD patterns of CdTe thin-films electrodeposited at different growth temperatures under as-deposited (AD) and cadmium chloride treated (CCT) conditions respectively. While Table 1; Table 2 are the respective summaries of the X-ray diffraction analysis for cubic (111) CdTe diffraction for AD and CCT-CdTe layers and the comparative analysis of the dislocation density and number of crystallites per unit area of the CdTe thin-films electroplated at a different temperature. Under both the AD and CCT conditions, the CdTe layers show diffraction patterns with a preferential and strong (111) phase of the cubic structure of CdTe at 2θ = ~24.0° (see Figure 3a,b). Diffractions with the underlying g/FTO substrate were observable at 2θ = ~20.6°, ~33.8°, ~37.9°, ~51.6°, ~60.7° and ~65.6°. This is in addition to the cadmium tellurate (Cd_xTeO_y) diffraction observed at 2θ = 23.0°. With respect to the AD-CdTe layers shown in Figure 3a and Table 1, an increase in the deposition temperature resulted in an increase in the cubic (111) CdTe diffraction intensity, peak sharpness, and crystallite size as evident by the 65.8 nm observed at 85 °C. This indicates an enhancement in the crystallinity and a reduction in the lattice defects of the CdTe-layers at high deposition temperature. This observation can also be said of the CCT-CdTe layers with the highest cubic (111) CdTe intensity at 85 °C (see Figure 3b and Table 1).

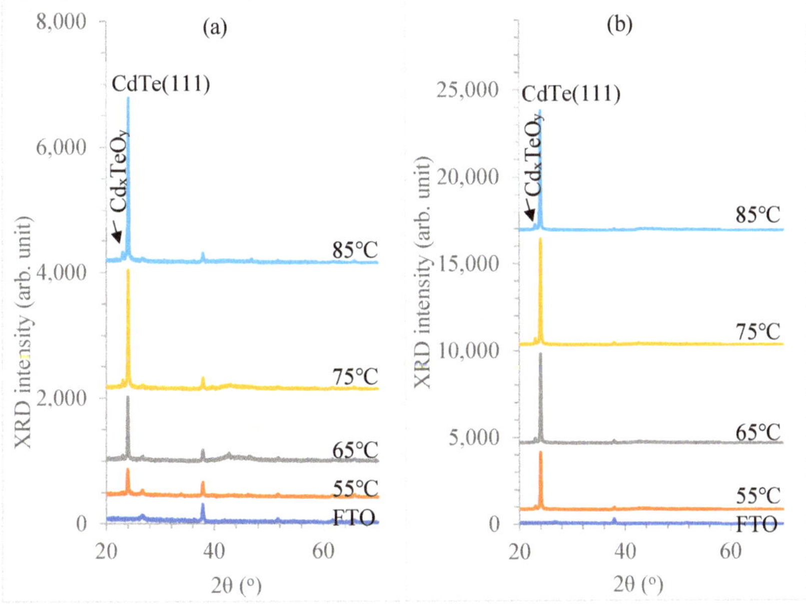

Figure 3. XRD patterns of CdTe thin-films electroplated at different temperatures under (**a**) as-deposited (AD) and (**b**) cadmium chloride treated (CCT) conditions.

With emphasis on the CdTe (111)C preferred orientation, a drastic improvement in the intensity of the diffraction after CCT treatment and crystallite size was observed notwithstanding the CdTe-bath deposition temperature. Interestingly, a reduction in the dislocation density and the number of crystallites per unit area was observed with an increase in the deposition temperature, while further improvements were observed even after CCT (see Table 2 and Figure 4). This observations might be as a result of the recrystallization of the polycrystalline CdTe thin-films during treatment which results to the improvement in the Cd/Te stoichiometry by the sublimation of surplus elements or the formation of CdTe by the reaction between surplus elemental Te and Cd from $CdCl_2$ treatment [4,21]. It should be noted that no matter the deposition technique, Te –precipitation is one of the main challenges in the deposition of CdTe [22,23]

Table 1. Summary of the X-ray diffraction analysis for cubic (111) CdTe diffraction for AD and CCT-CdTe layers.

Deposition Temperature ($\pm 2\,^\circ$C)	2θ ($^\circ$)	Lattice Spacing (Å)	FWHM ($^\circ$)	Crystallite Size D (nm)	Dislocation Density $\delta_1 \times 10^{11}$ (Lines·cm^{-2})	Number of Crystallites per Unit Area N_1 $\times 10^{12}$ (cm^{-2})
			AD			
55	23.9	3.73	0.227	37.4	7.15	1.91
65	23.8	3.74	0.195	43.5	5.28	1.21
75	23.9	3.72	0.162	52.4	3.64	0.69
85	24.0	3.70	0.129	65.8	2.31	0.35
			CCT			
55	23.9	3.714	0.162	52.4	3.64	0.69
65	24.0	3.708	0.129	65.8	2.31	0.35
75	24.0	3.713	0.129	65.8	2.31	0.35
85	24.0	3.715	0.129	65.8	2.31	0.35

Improvement in the crystallinity of CdTe after post-growth treatment in the presence of chlorine due to factors such as grain growth, recrystallization amongst others is generally accepted by the scientific community [24,25]. The stagnation observed in the crystallite size at 65.8 nm might be as a result of the limitation of the XRD machine utilized and/or that of the Scherrer's formula utilized in the measurement and analysis of CdTe thin-films respectively [26,27].

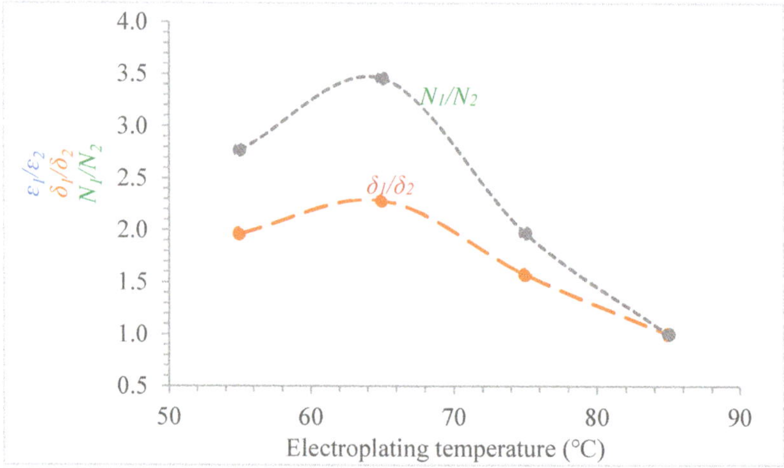

Figure 4. Comparative analysis of the dislocation density and number of crystallites per unit area of the CdTe films electroplated at different cathodic voltages.

Table 2. Comparative analysis of the dislocation density and number of crystallites per unit area of the CdTe films electroplated at different temperature under AD and CCT conditions.

Deposition Temperature ($\pm 2\,^{\circ}$C)	δ_1/δ_2	N_1/N_2
55	1.96	2.77
65	2.29	3.46
75	1.58	1.97
85	1.00	1.00

The obtained XRD diffraction data from the CdTe thin film structural analysis is in agreement with the 01-075-2086 reference file of the Joint Committee on Powder Diffraction Standards (JCPDS) for cubic CdTe layers.

3.2. Thickness Measurement

For this experiment, the deposition duration was maintained at 180 min for each of the electroplated CdTe at different deposition temperatures. Figure 5 shows the graphical plot of both the estimated and the measured thicknesses including the average deposition current density of the CdTe layers against the deposition temperature. For the estimated thickness using the Faraday's equation (see Equation (4)), the rise in the thickness of electrodeposited CdTe with increasing deposition temperature was solely due to the corresponding increase in the average current density of deposition. In addition, increase in the solubility of solvents and the catalyses, the reactions are also as a result of an increase in the electroplating temperature [28]. Correspondingly, this divulges an increase in the deposition current density and hence, a rapid growth rate of constituent elements or compounds.

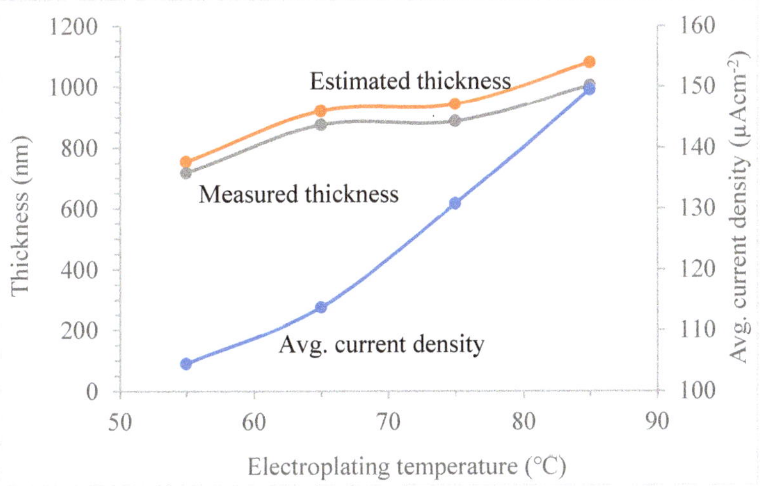

Figure 5. A plot of the estimated and measured thicknesses (as-deposited) and the average current density during electroplating against the electroplating temperature for CdTe layers grown for 180 min.

The observed deviation of the estimated thickness and the measured thickness is due to the assumption made by Faraday's law of electrolysis that all the electronic charges flowing through the electrolytic cell partake in the growth of the electroplated materials. The assumption did not consider the electronic charges associated with the breakdown of water into its constituent ions and the additional chemical reactions on both the cathode and anode. Therefore, the estimated thickness using Faraday's formula served as the upper limit of the thickness. As observed in Figure 5, a greater deviation between the measured and the estimated thickness was observed for the CdTe layers grown at 85 °C due to the favourable impact of higher temperature to molecular and ionic activity within the electrolyte.

3.3. Optical Properties Analysis

Figure 6a,b shows the Tauc's plot of $(\alpha h v)^2$ against $(h v)$ for CdTe thin-films electrodeposited at different temperatures under both AD and CCT conditions respectively. Table 3 and Figure 7 summarizes the observed bandgap energy and the absorption edge of the investigated CdTe thin-films. Under the as-deposited condition, the observed bandgaps range is within the generally acceptable range of 1.51 ± 0.03 eV for CdTe layers [29]. This follows a trend in which the highest bandgap of 1.54 eV was recorded at 55 °C and the lowest bandgap of 1.48 eV was observed at 85 °C. It is generally accepted in the photovoltaic community that the optimal bandgap 1.45 eV is required to achieve the highest efficiency of a one-bandgap *p-n* junction photovoltaic device.

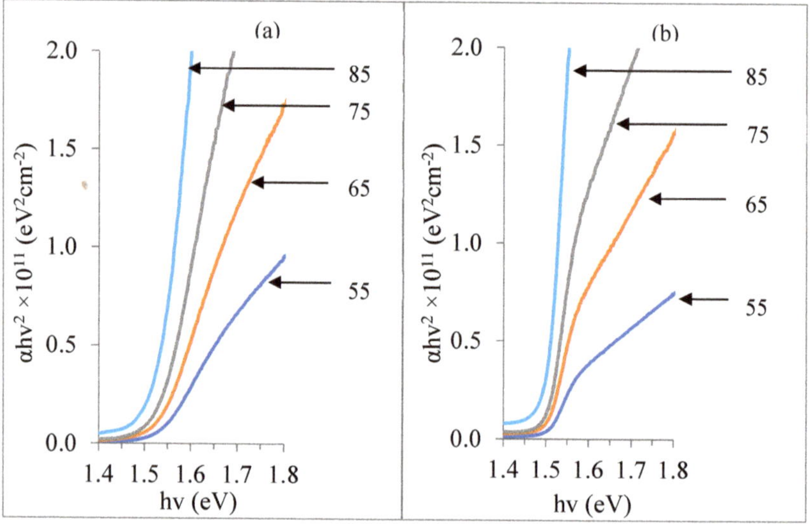

Figure 6. Optical absorption spectra for electroplated CdTe thin-films electroplated at different deposition temperature under (**a**) AD and (**b**) CCT conditions.

Table 3. The optical bandgap and absorption edge slope of CdTe layers electroplated from electrolytes at different temperatures.

Electroplating Temperature (°C)	Bandgap (eV)		Absorbance Edge Slope (eV^{-1})	
	AD	CCT	AD	CCT
55	1.54	1.50	4.17	4.34
65	1.53	1.49	6.67	8.33
75	1.52	1.49	8.33	9.09
85	1.48	1.46	9.09	10.00

A shift towards 1.45 eV bandgap was observed after post-growth treatment –CCT as shown in Figure 6b, Figure 7 and Table 3. This observation should be as a result of the improvement in the material and electronic properties of CdTe [24,25]. Further to this, a shift in the absorption edge slope was also observable, with the highest edge slope observed at 85 °C under both AD and CCT conditions. It is known that the sharpness of the absorption edge slope indicates lesser impurity energy levels and defects in the thin-films under investigation [30–33]. This observation suggests that more stoichiometric CdTe layers are deposited at 85 °C and a further improvement is achievable after CCT. This observation is in agreement with the summation made in Section 3.1.

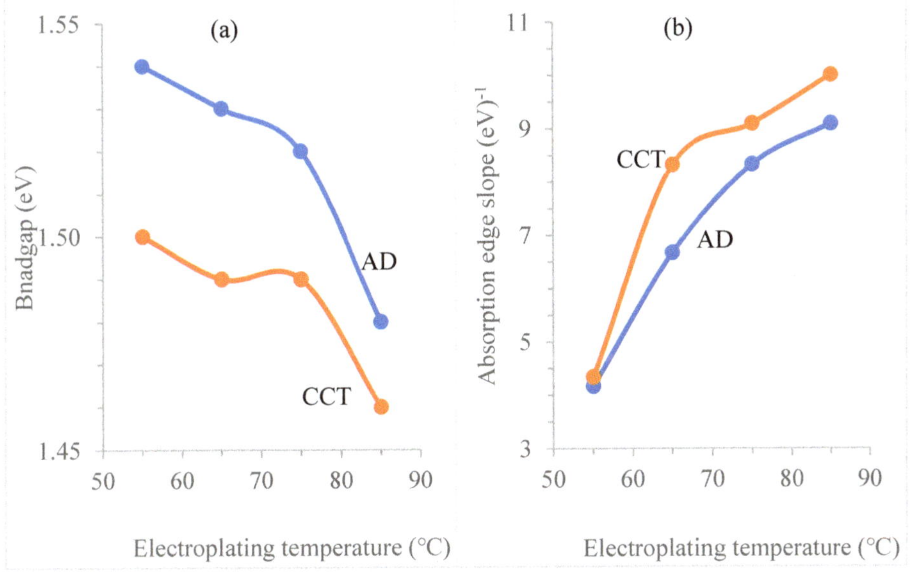

Figure 7. (**a**) A plot of optical bandgap of CdTe electroplated from electrolytes at different deposition temperature and (**b**) is the absorption edge slope of the CdTe layers under both AD and CCT conditions against deposition temperature.

3.4. Morphological Properties Analysis

Figure 8a–d shows the SEM images of CCT-CdTe thin-films electrodeposited at different growth temperatures ranging from 55 to 85 °C. The CdTe thin-films deposited at all the explored deposition temperatures show excellent coverage of the underlying g/FTO substrate. The purported good coverage is due to the good quality of the CdTe layers deposited based on the electrodeposition parameters such as the cathodic voltage, stirring rate, pH amongst other factors prior to CCT. While the retention of the complete coverage after CCT is owing to the enhancements in the CdTe micro-structure as a result of the recrystallization, sublimation of superfluous elements, grain growth and the formation of CdTe via a chemical reaction between excess Cd from $CdCl_2$ and precipitated Te in the layer. In AD-CdTe, cauliflower-like clusters consisting of numerous crystals are often observed with sizes ranging between ~30 and 65 nm [17]. Upon CCT, these crystals merge into grains due to their large surface to volume ratio. This is a key feature of nano-materials and helps in the formation of large grains. The CCT-CdTe grains are much larger with grains sizes ranging between ~200 and 6000 nm.

Based on topological observation, a gradual increase in the grain size with increasing deposition temperature was noted with the largest grain size of 6 µm and lower grain boundary density observed for the 85 °C bath. This observation indicates that although CCT facilitates grain growth in CdTe, high deposition temperature favours grain growth. It is relevant at this point to compare the noted trend of the grain size in Figure 8 with the deductions made based on the number of crystallites per unit area N_1 in Section 3.1. It is at this point important to note that the calculated crystallite size using the Scherrer formula in XRD analysis does not translate to the sizes of grains observed in the SEM micrograph. However, grains can be formed from a single or multiple crystallites.

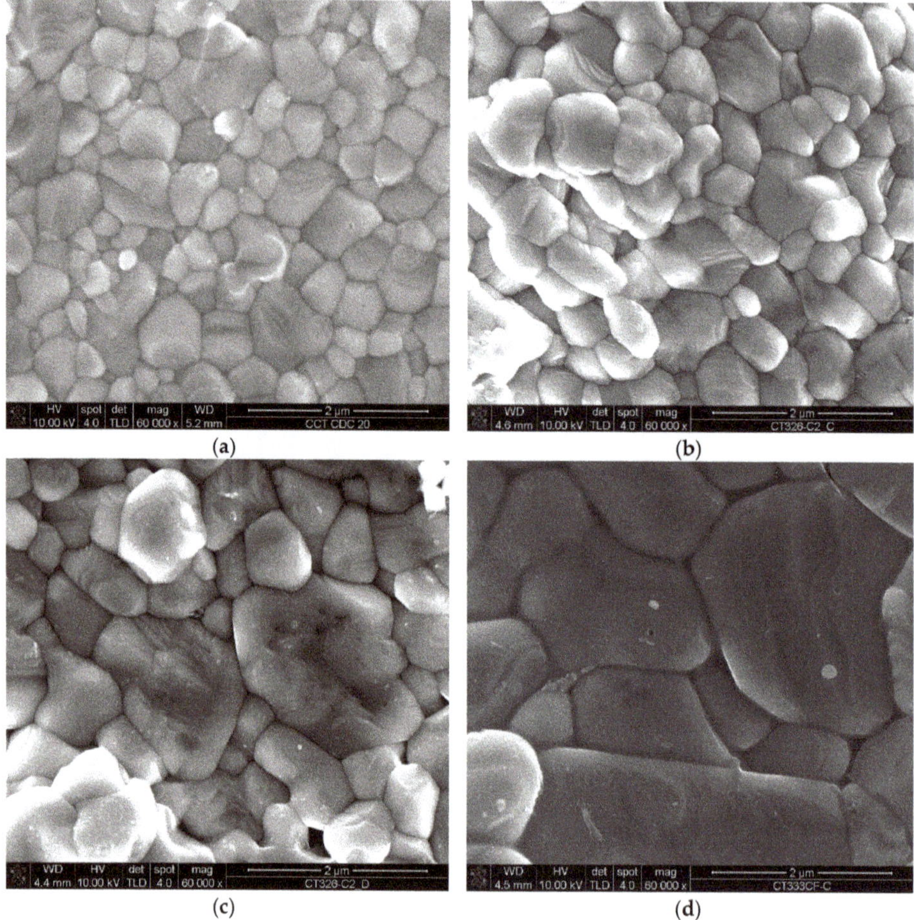

Figure 8. Typical 2-dimensional SEM micrographs for CdCl$_2$ treated CdTe samples grown from electrolyte at different electroplating temperature between 55 and 85 °C. (**a**) 55 °C CCT-CdTe; (**b**) 65°C CCT-CdTe; (**c**) 75 °C CCT-CdTe; (**d**) 85 °C CCT-CdTe.

3.5. Compositional Measurement

Figure 9 shows the typical EDX spectra for AD and CCT-CdTe grain at 85 °C and the plot of the atomic composition ratio of Te to Cd in electroplated CdTe layers under both AD and CCT conditions against the electroplating temperature. Aside from Te and Cd elements, other elements such as O, F, Sn, Si were also randomly observed owing to layer oxidation and/or the underlying g/FTO substrate. With emphasis on the as-deposited CdTe layers as shown in Figure 9, the compositional ratio of Te/Cd is less than 1 for the CdTe thin-film deposited at 85 °C. The CdTe thin-films grown at deposition temperature 75 °C and below show comparative Te-richness.

It can be recalled from the succinct discussion on the cyclic voltammogram in Section 2.1 and Figure 2 that with the increasing deposition potential, electroplated layers go through the stages of Te deposition, Te-rich CdTe deposition, near stoichiometric CdTe deposition and Cd-rich CdTe depositions based on the redox potential of Cd and Te. The observed effect of deposition temperature even when it is performed at the same deposition potential seems to give a similar effect by a continued shift in the cyclic voltammetric graph to the right with decreasing deposition temperature. This argument is valid due to the catalytic effect of heat in reaction and in the mobility of ions constituted in the electrolyte especially Cd with comparatively lower redox potential. This is one of the possible reasons for the Cd or Te richness of the deposited CdTe layer at the same cathodic voltage at the exploring temperature of 85 °C and between 55 and 85 °C, respectively. After CCT, a shift towards stoichiometry in the atomic composition ratio of the CdTe layers was observed (see Figure 9). This might be influenced by the reaction between Cd from $CdCl_2$ with unreacted Te and/or the sublimation of excess elemental Cd and Te from the layer. For photovoltaic applications, the richness of Cd in CdTe used as an absorber layer has been documented in the literature to produce a comparatively higher efficiency [34–37]. A stoichiometry of 50/50 Te to Cd atomic composition was observed for the 85 °C CdTe after CCT signify comparatively higher crystallinity. This observation is in agreement with the high crystallinity level observed at 85 °C.

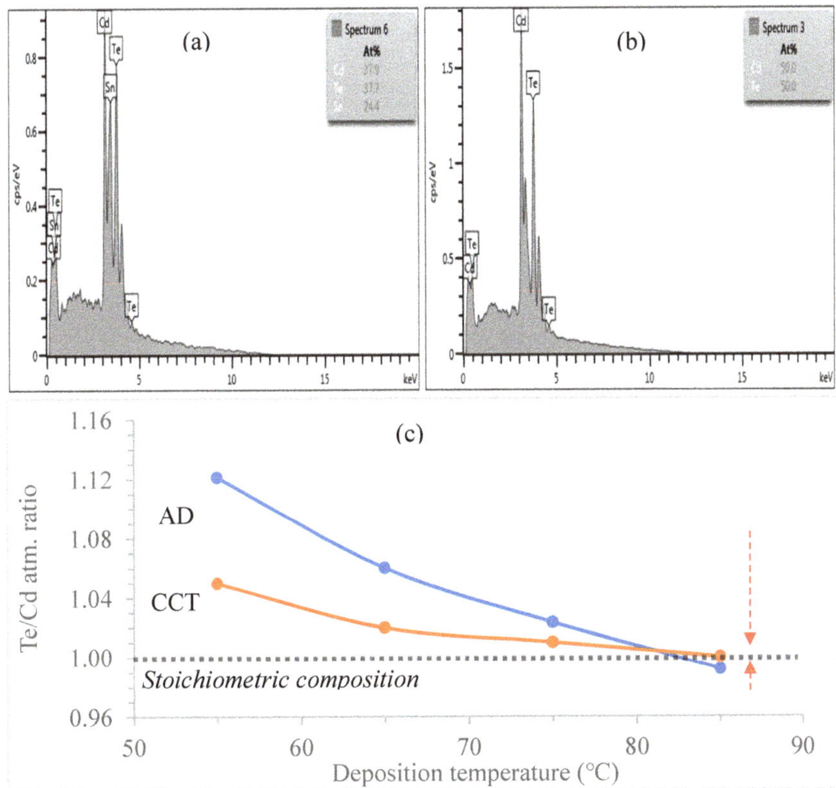

Figure 9. EDX spectra of CdTe grown from the 85 °C electrolyte under (**a**) AD, (**b**) CCT conditions (**c**) is the graphical summary of Te/Cd atomic composition ratio against electroplating temperature of CdTe for both AD and CCT samples.

3.6. Photoelectrochemical (PEC) Cell Measurement

Figure 10 shows the PEC cell measurements of the CdTe layers electroplated from electrolytes at different deposition temperature against electroplating temperature under both AD and CCT conditions. An all-round technique such as the Hall effect measurement was not used because of the effect of the highly conductive FTO underlying substrate. The obtained data cannot be isolated only for CdTe from the measurement.

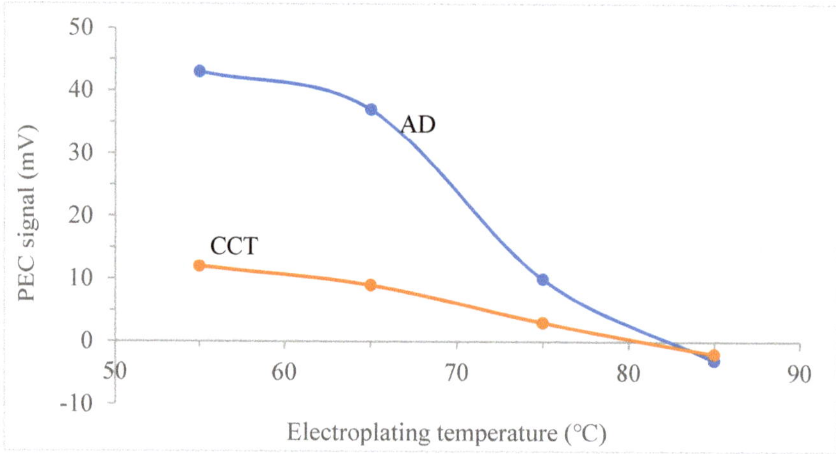

Figure 10. A plot of PEC signal of CdTe electroplated from electrolytes at different deposition temperatures against electroplating temperature under both AD and CCT conditions.

As depicted in Figure 10, the electrical conduction type of the as-deposited CdTe layers grown at 85 °C is *n*-type while the ones grown at between 55 and 75 °C, are *p*-type. A similar observation is also made of the CdTe layers after CCT with a shift from the *n*-type layers towards p-type electrical conductivity and vice versa, while the initial conduction types were retained. The observed PEC signal shift after CCT is an attribute of recrystallization amongst factors that have been well document [24,25]. It is captivating that the observed PEC trend is somewhat similar to that of the EDX spectra summary in Figure 9c. This is because one of the predominant factors determining the electrical conduction type is the atomic compositional ratio of the constituent elements, with Te-richness in CdTe resulting in p-type and Cd-richness resulting in *n*-type conduction type [18,34,35]. Other factors include the heat treatment parameters (such as duration, temperature), preliminary atomic composition of Cd and Te, the concentration of $CdCl_2$ or other dopants utilized in treatment, the structure of the defects in the material, and the initial conductivity type of the material [10,17,24,38].

4. Conclusions

The effect of deposition temperature in a two-electrode electrodeposition configuration was explored and the ensued CdTe thin-films were methodically characterized and presented. The structural, optical, morphological, compositional, and electronic properties of the CdTe relative to their deposition temperature were studied. All the electroplated thin-films grown at a different deposition temperature of the electrolyte show polycrystalline cubic structure of the material with a preferred orientation along the (111) plane. The dislocation density for the as-deposited CdTe was observed to be 7.15×10^{11} lines·cm^{-2} for the CdTe deposited at 55 °C and 2.31×10^{11} lines·cm^{-2} for the grown CdTe at 85 °C, which was respectively reduced to 3.64 and 2.31 lines·cm^{-2}. The number of crystallites per unit area N was found to reduce from 1.91×10^{12} to 0.35×10^{12} cm^{-2} with increased deposition temperature from 55 to 85 °C. The N was found to reduce to 0.69×10^{12} to 0.35×10^{12} cm^{-2} with increased deposition temperature from 55 to 85 °C after CCT. This is an indication that the crystallite sizes increases after annealing. This was evident with the crystallite sizes of the as-deposited films ranging from 37.4 to 65.8 nm with increased deposition temperature from 55 to 85 °C to 52.4 to 65.8 nm after post-growth CCT. The optical property investigation reveals that, the deposited layers possess bandgaps ranging between 1.51 ± 0.03 eV under as-deposited condition and 1.48 ± 0.02 eV after CCT. Prominently, the sharpness of the optical absorption edge slope reduces with the reduction in the deposition temperature. Morphologically, all the electrodeposited layers show full underlying layer coverage.

Comparatively, larger grains after CCT were observed for layers grown at 85 °C. The compositional analysis reveals the presence of Cd and Te in the deposited thin-film. A atomic ratio for Cd:Te of 1:1 was recorded for CdTe layers grown at an electroplating temperature of 85 °C. Additionally, an increase in the atomic concentration of Te with the reduction of the electroplating temperature for the explore deposition temperature range of 85 to 55 °C was noted. The PEC measurements show that the CdTe layers grown at 85 °C is *n*-type under both AD and CCT conditions, while p-type conduction type CdTe layers ensued for the layers grown at deposition temperature of 75 °C and below. These results underline the importance of the deposition temperature in the electrodeposition and the capability of two-electrode electrodeposition configuration. Aside from the elimination of possible contaminants from the reference electrode in this configuration, the 2-electrode system also provides the leeway of electroplating at higher temperatures to improve the material and electronic qualities of PV materials.

Author Contributions: The individual contributions to this publication include the conceptualization, A.A.O.; methodology, A.A.O.; validation, A.A.O. and I.M.D.; formal analysis, A.A.O.; investigation, A.A.O.; resources, A.A.O. and I.M.D.; data curation, A.A.O.; writing—original draft preparation, A.A.O.; writing—review and editing, A.A.O.; visualization, I.M.D.; supervision, I.M.D.; project administration, I.M.D.

Funding: This research received no external funding.

Acknowledgments: The main author would like to profoundly appreciate the Materials and Engineering Research Institute (MERI), Sheffield Hallam University (SHU) for the equipment utilised for material characterisation. In addition, the University of Ado Ekiti is also acknowledged for their moral support.

Conflicts of Interest: The authors declare no conflict of interest.

References

1. Lincot, D. Electrodeposition of semiconductors. *Thin Solid Films* **2005**, *487*, 40–48. [CrossRef]
2. Dharmadasa, I.M.; Madugu, M.L.; Olusola, O.I.; Echendu, O.K.; Fauzi, F.; Diso, D.G.; Weerasinghe, A.R.; Druffel, T.; Dharmadasa, R.; Lavery, B.; et al. Electroplating of CdTe thin films from cadmium sulphate precursor and comparison of layers grown by 3-electrode and 2-electrode systems. *Coatings* **2017**, *7*, 17. [CrossRef]
3. Ojo, A.A.; Dharmadasa, I.M. Electroplating of semiconductor materials for applications in large area electronics: A review. *Coatings* **2018**, *8*, 262. [CrossRef]
4. Bosio, A.; Romeo, N.; Mazzamuto, S.; Canevari, V. Polycrystalline CdTe thin films for photovoltaic applications. *Prog. Cryst. Growth Charact. Mater.* **2006**, *52*, 247–279. [CrossRef]
5. Turner, A.; Woodcock, J.; Ozsan, M.; Cunningham, D.; Johnson, D.; Marshall, R.; Mason, N.; Oktik, S.; Patterson, M.; Ransome, S. BP solar thin film CdTe photovoltaic technology. *Sol. Energy Mater. Sol. Cells* **1994**, *35*, 263–270. [CrossRef]
6. McCandless, B.E.; Sites, J.R. Cadmium Telluride Solar Cells. In *Handbook of Photovoltaic Science and Engineering*; John Wiley & Sons, Ltd.: Chichester, UK, 2005; pp. 617–662.
7. Green, M.A.; Hishikawa, Y.; Dunlop, E.D.; Levi, D.H.; Hohl-Ebinger, J.; Ho-Baillie, A.W.Y. Solar cell efficiency tables (version 51). *Prog. Photovolt. Res. Appl.* **2018**, *26*, 3–12. [CrossRef]
8. Wu, X. High-efficiency polycrystalline CdTe thin-film solar cells. *Sol. Energy* **2004**, *77*, 803–814. [CrossRef]
9. First Solar Raises Bar for CdTe with 21.5% Efficiency Record. *pv-magazine.* (n.d.) Available online: http://www.pv-magazine.com/news/details/beitrag/first-solar-raises-bar-for-cdte-with-215-efficiency-record_100018069/#axzz3rzMESjUl (accessed on 9 April 2019).
10. Abdul-Manaf, N.A.; Salim, H.I.; Madugu, M.L.; Olusola, O.I.; Dharmadasa, I.M. Electro-plating and characterisation of CdTe thin films using CdCl$_2$ as the cadmium source. *Energies* **2015**, *8*, 10883–10903. [CrossRef]
11. Echendu, O.K.; Okeoma, K.B.; Oriaku, C.I.; Dharmadasa, I.M. Electrochemical deposition of CdTe semiconductor thin films for solar cell application using two-electrode and three-electrode configurations: A comparative study. *Adv. Mater. Sci. Eng.* **2016**, *2016*, 3581725. [CrossRef]
12. Ojo, A.A.; Dharmadasa, I.M. Analysis of electrodeposited CdTe thin films grown using cadmium chloride precursor for applications in solar cells. *J. Mater. Sci. Mater. Electron.* **2017**, *28*, 14110–14120. [CrossRef]

13. Gómez, H.; Henríquez, R.; Schrebler, R.; Córdova, R.; Ramírez, D.; Riveros, G.; Dalchiele, E.A. Electrodeposition of CdTe thin films onto n-Si(100): Nucleation and growth mechanisms. *Electrochim. Acta* **2005**, *50*, 1299–1305. [CrossRef]

14. Murase, K.; Honda, T.; Yamamoto, M.; Hirato, T.; Awakura, Y. Electrodeposition of CdTe from basic aqueous solutions containing ethylenediamine. *J. Electrochem. Soc.* **2001**, *148*, 203. [CrossRef]

15. Das, S.K.; Morris, G.C. Influence of growth and microstructure of electrodeposited cadmium telluride films on the properties of n-CdS/p-CdTe thin-film solar cells. *J. Appl. Phys.* **1992**, *72*, 4940–4945. [CrossRef]

16. Kumarasinghe, K.D.M.S.P.K.; De Silva, D.S.M.; Pathiratne, K.A.S.; Salim, H.I.; Abdul-Manaf, N.A.; Dharmadasa, I.M. Electrodeposition and characterization of as-deposited and annealed CdTe thin films. *Ceylon J. Sci.* **2016**, *45*, 53. [CrossRef]

17. Salim, H.I.; Patel, V.; Abbas, A.; Walls, J.M.; Dharmadasa, I.M. Electrodeposition of CdTe thin films using nitrate precursor for applications in solar cells. *J. Mater. Sci. Mater. Electron.* **2015**, *26*, 3119–3128. [CrossRef]

18. Ojo, A.A.; Cranton, W.M.; Dharmadasa, I.M. *Next Generation Multilayer Graded Bandgap Solar Cells*; Springer: Cham, Switzerland, 2019.

19. Williamson, G.K.; Smallman, R.E., III. Dislocation densities in some annealed and cold-worked metals from measurements on the X-ray debye-scherrer spectrum. *Philos. Mag.* **1956**, *1*, 34–46. [CrossRef]

20. Prabahar, S.; Dhanam, M. CdS thin films from two different chemical baths—Structural and optical analysis. *J. Cryst. Growth* **2005**, *285*, 41–48. [CrossRef]

21. Dharmadasa, I.M.; Echendu, O.K.; Fauzi, F.; Abdul-Manaf, N.A.; Olusola, O.I.; Salim, H.I.; Madugu, M.L.; Ojo, A.A. Improvement of composition of CdTe thin films during heat treatment in the presence of $CdCl_2$. *J. Mater. Sci. Mater. Electron.* **2017**, *28*, 2343–2352. [CrossRef]

22. Tranchart, J.; Bach, P. A gas bearing system for the growth of CdTe. *J. Cryst. Growth* **1976**, *32*, 8–12. [CrossRef]

23. Sochinskii, N.; Babentsov, V.; Tarbaev, N.; Serrano, M.; Diéguez, E. The low temperature annealing of p-cadmium telluride in gallium-bath. *Mater. Res. Bull.* **1993**, *28*, 1061–1066. [CrossRef]

24. Dharmadasa, I.M. Review of the $CdCl_2$ treatment used in CdS/CdTe thin film solar cell development and new evidence towards improved understanding. *Coatings* **2014**, *4*, 282–307. [CrossRef]

25. Bosio, A.; Rosa, G.; Menossi, D.; Romeo, N. How the chlorine treatment and the stoichiometry influences the grain boundary passivation in polycrystalline CdTe thin films. *Energies* **2016**, *9*, 254. [CrossRef]

26. Monshi, A.; Foroughi, M.R. Modified Scherrer Equation to Estimate More Accurately Nano-Crystallite Size Using XRD. *World J. Nano Sci. Eng.* **2012**, *2*, 154–160. [CrossRef]

27. Bouraiou, A.; Aida, M.; Meglali, O.; Attaf, N. Potential effect on the properties of $CuInSe_2$ thin films deposited using two-electrode system. *Curr. Appl. Phys.* **2011**, *11*, 1173–1178. [CrossRef]

28. Rakhshani, A.E.; Varghese, J. The effect of temperature on electrodeposition of cuprous oxide. *Phys. Status Solidi A* **1988**, *105*, 183–188. [CrossRef]

29. Sze, S.; Ng, K.K. *Physics of Semiconductor Devices*; John; Wiley & Sons, Inc.: Hoboken, NJ, USA, 2006.

30. Han, J.; Spanheimer, C.; Haindl, G.; Fu, G.; Krishnakumar, V.; Schaffner, J.; Fan, C.; Zhao, K.; Klein, A.; Jaegermann, W. Optimized chemical bath deposited CdS layers for the improvement of CdTe solar cells. *Sol. Energy Mater. Sol. Cells* **2011**, *95*, 816–820. [CrossRef]

31. Marple, D.T.F. Optical Absorption Edge in CdTe: Experimental. *Phys. Rev.* **1966**, *150*, 728–734. [CrossRef]

32. Pérez-Hernández, G.; Pantoja-Enríquez, J.; Escobar-Morales, B.; Martínez-Hernández, D.; Díaz-Flores, L.; Ricardez-Jiménez, C.; Mathews, N.; Mathew, X. A comparative study of CdS thin films deposited by different techniques. *Thin Solid Films* **2013**, *535*, 154–157. [CrossRef]

33. Metin, H.; Esen, R. Annealing effects on optical and crystallographic properties of CBD grown CdS films. *Semicond. Sci. Technol.* **2003**, *18*, 647–654. [CrossRef]

34. Reese, M.O.; Perkins, C.L.; Burst, J.M.; Farrell, S.; Barnes, T.M.; Johnston, S.W.; Kuciauskas, D.; Gessert, T.A.; Metzger, W.K. Intrinsic surface passivation of CdTe. *J. Appl. Phys.* **2015**, *118*, 155305. [CrossRef]

35. Burst, J.M.; Duenow, J.N.; Albin, D.S.; Colegrove, E.; Reese, M.O.; Aguiar, J.A.; Jiang, C.-S.; Patel, M.K.; Al-Jassim, M.M.; Kuciauskas, D.; et al. CdTe solar cells with open-circuit voltage breaking the 1 V barrier. *Nat. Energy* **2016**, *1*, 16015. [CrossRef]

36. Diso, D.G.; Fauzi, F.; Echendu, O.K.; Olusola, O.I.; Dharmadasa, I.M. Optimisation of CdTe electrodeposition voltage for development of CdS/CdTe solar cells. *J. Mater. Sci. Mater. Electron.* **2016**, *27*, 12464–12472. [CrossRef]

37. Dharmadasa, I.M.; Ojo, A.A. Unravelling complex nature of CdS/CdTe based thin film solar cells. *J. Mater. Sci. Mater. Electron.* **2017**, *28*, 16598–16617. [CrossRef]
38. Başol, B.M. Processing high efficiency CdTe solar cells. *Int. J. Sol. Energy* **1992**, *12*, 25–35. [CrossRef]

Article

A Deep Insight into the Electronic Properties of CIGS Modules with Monolithic Interconnects Based on 2D Simulations with TCAD

Ricardo Vidal Lorbada [1,2,*], Thomas Walter [2], David Fuertes Marrón [1], Tetiana Lavrenko [2] and Dennis Muecke [2]

[1] Instituto de Energía Solar, Universidad Politécnica de Madrid, 28040 Madrid, Spain; dfuertes@ies.upm.es
[2] Institute of Medical Engineering and Mechatronics, Ulm University of Applied Sciences, 89081 Ulm, Germany; walter.th@hs-ulm.de (T.W.); lavrenko@hs-ulm.de (T.L.); muecke@mail.hs-ulm.de (D.M.)
* Correspondence: vidal@hs-ulm.de; Tel.: +34-626-152-699

Received: 30 January 2019; Accepted: 17 February 2019; Published: 19 February 2019

Abstract: The aim of this work is to provide an insight into the impact of the P1 shunt on the performance of $ZnO/CdS/Cu(In,Ga)Se_2/Mo$ modules with monolithic interconnects. The P1 scribe is a pattern that separates the back contact of two adjacent cells and is filled with $Cu(In,Ga)Se_2$ (CIGS). This scribe introduces a shunt that can affect significantly the behavior of the device, especially under weak light conditions. Based on 2D numerical simulations performed with TCAD, we postulate a mechanism that affects the current flow through the P1 shunt. This mechanism is similar to that of a junction field effect transistor device with a p-type channel, in which the current flow can be modulated by varying the thickness of the channel and the doping concentration. The results of these simulations suggest that expanding the space charge region (SCR) into P1 reduces the shunt conductance in this path significantly, thus decreasing the current flow through it. The presented simulations demonstrate that two fabrication parameters have a direct influence on the extension of the SCR, which are the thickness of the absorber layer and its acceptor concentration.

Keywords: $Cu(In,Ga)Se_2$; mini-module; numerical simulation; P1 shunt; space charge region (SCR); TCAD; transistor effect

---

1. Introduction

The prospects of copper indium gallium diselenide-based solar cells (CIGS) as an alternative to traditional silicon solutions are growing with each efficiency achievement. Three devices are distinguished regarding their respective efficiencies, which are lab-scale, mini-modules, and modules. Regarding lab scale devices, ZSW reported an efficiency of 22.6% [1], meanwhile Solar Frontier reported the world record with an efficiency of 22.9% [2]. In the case of mini-module devices, 18.7% efficiency was reported by Solibro [3]. Finally, TSMC reported an efficiency of 16.5% for a full-size module in reference [4].

According to reference [5], improving the efficiency of lab scale cells could show the potential for improvement of large area interconnected devices. Optimization of ZnO window layers, absorbers and CIGS/CdS interfaces are found to be a key for the improvement of laboratory size devices [5]. On the other hand, according to reference [4], taking into consideration the dead areas of interconnects could also reduce power losses and improve module efficiency.

In this work, we present simulations of the dead area in interconnected devices with varied fabrication parameters to study their impact on the dark *JV*-characteristic. We will also show experimental results to compare them with the simulations. Furthermore, based on these simulations and experimental results we will explain a mechanism that plays an important role in the JV-characteristic of the device.

Thin film CIGS devices consist of five functional layers [6,7]. The monolithic interconnects between two devices are introduced through three scribes [7,8]. This structure is known to induce some efficiency losses due to shunts and dead areas [8,9]. In this work we investigate the shunt associated with the P1 scribe between adjacent molybdenum contacts. It has a direct impact on the JV-characteristic of the device, as demonstrated in reference [8], and is related both to the resistivity of the material and the width of the scribe. The scribe width can be adjusted easily to increase the resistance between both molybdenum contacts and reduce the P1 shunt; however, it also increases the dead area, and according to reference [8] there is a superior limit in which the efficiency starts to decrease. The schematic given in Figure 1 represents the target structure for simulation, adapted from [7] but simplified for the scope of this work.

Figure 1. Schematic of Cu(In,Ga)Se$_2$ (CIGS) device with ZnO/CdS/CIGS/Moly and P1, P2, P3 scribes.

The mechanism discussed in this contribution is closely related to the working principle of a junction field effect transistor device (JFET), similar to the simplified p-type channel model with three layers presented in reference [10]. On top of the JFET device there is an n-type semiconductor layer with a gate contact placed over it. Under this layer there is a p-type semiconductor, which corresponds to the channel, being the main current path. The model exhibits two lateral contacts, the source, and the drain.

A space charge region (SCR) is created in the region near the p–n junction. The width of this SCR can be controlled by applying a bias to the gate or adjusting the values of acceptor (N_A) and donor (N_D) concentrations for the p-type and n-type semiconductor layers, respectively. A negative bias applied to the gate will increase the width of the SCR, whereas a positive bias will reduce it. The applied bias modulates the width of the SCR at the expense of the channel; thus, the current level in the channel is adjusted.

In the case of CIGS devices with monolithic interconnects the situation is similar. The top n-type semiconductor is ZnO/CdS. The intermediate p-type CIGS layer, including the P1 scribe, corresponds to the channel, which is placed between both molybdenum contacts. These contacts correspond to the source and the drain in the previous model. There is also a SCR in the junction between the ZnO/CdS n-type and the CIGS p-type semiconductor.

Numerical simulation is a powerful tool to investigate and optimize CIGS in different contexts. In reference [11], 2D simulation models of CIGS devices with monolithic interconnects are presented. The impact of different parameter variations, including the width of the P1 scribe and the value of N_A on the performance ratio and low light behavior are also studied in reference [11]. Another contribution that deals with 2D numerical simulations of CIGS devices is reference [12], in which passivation of the CIGS/CdS interface is studied, although the presented model does not include the interconnect. The numerical optimization of the width of the point-contact used for passivation of the CIGS/CdS interface is possible according to reference [12].

Numerical simulations can also be used to study new materials for CIGS devices. In reference [13], bandgap gradient optimization of CIGS devices with ZnS buffer layer was performed. The main result presented in reference [13] is how the bandgap gradient in CIGS affects the carrier transportation, and in turn the device performance.

Finally, simulation is the most efficient method for device optimization before production. The work presented in reference [14] shows how the thickness of the CIGS and ZnO layers can be adjusted to increase the efficiency of CIGS thin-film devices.

Our contribution shows how these numerical simulation tools could be used to propose new models that explain the behavior of large area CIGS solar cells.

2. Materials and Methods

Simulations—Our first model, which is presented in Figure 2, replicates an element of a module with a monolithic interconnect, including a P1 and P2 scribe. The P3 scribe is not included in our model as it does not add any significant contribution to the discussed mechanism. The simulation is performed using Sentaurus TCAD (Synopsys, Mountain View, CA, USA) [15], which solves the Poisson and continuity equation sets for both electrons and holes [16,17]. The electronic and physical properties, as well as the physical dimensions of the device are presented in Table 1, which includes the geometry of the P1 and P2 scribes and total dead area. The temperature for all simulations is set at 300 K.

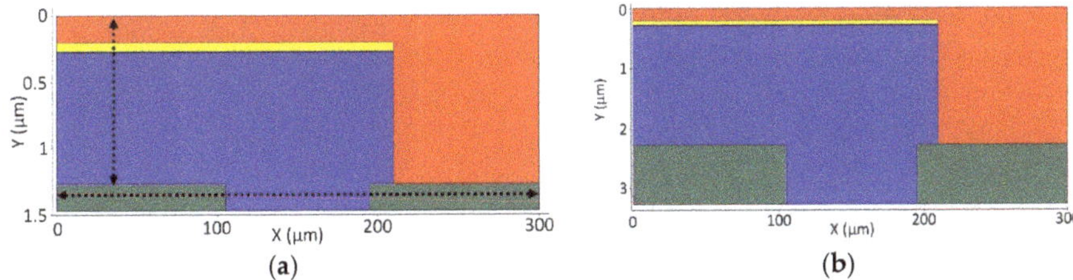

Figure 2. (**a**) Device structure with reduced CIGS and molybdenum thickness (1 and 0.2 μm, respectively), the vertical and horizontal lines indicate the positions of the vertical and horizontal cuts of 2D results, respectively; (**b**) device structure with extended CIGS and molybdenum thickness (2 and 1 μm, respectively).

Table 1. Key parameters of the functional layers with the dimensions of scribes and dead area.

Layer	N_D (cm^{-3})	N_A (cm^{-3})	X (eV)	μ_e (cm^2/V·s)	μ_h (cm^2/V·s)	E_g (eV)	Thickness (μm)	Scribe	Width (μm)
ZnO	10^{20}	–	4.6	50	20	3.5	0.2	P1	90
CdS	10^{17}	–	4.4	50	20	2.5	0.05	P2	90
CIGS	–	10^{14}–10^{16}	4.6	50	20	1.2	1.00–2.00	Dead area	200
Moly	–	–	–	–	–	–	0.20–1.0	Total	300

The parameters presented in Table 1 are in the same range as those seen in the literature [11,12]. The values of the electron and hole mobilities (μ_e and μ_h, respectively) could be considered overestimated, although in the contributions presented previously [12,13], the values of the mobilities for both carrier types are even higher (μ_e = 100 and μ_h = 25 cm^2/V·s).

Figure 3 introduces a second model that is relevant for the scope of this work. The aim of this simulation is to verify whether the presence of the space charge region (SCR) has an impact on the current flowing between anode and cathode through the P1-shunt. As such, the presented model has only a CIGS layer over the patterned (P1 scribe) molybdenum layer presented previously. Without the ZnO/CdS n-type semiconductor layers there is no junction, therefore the SCR is eliminated.

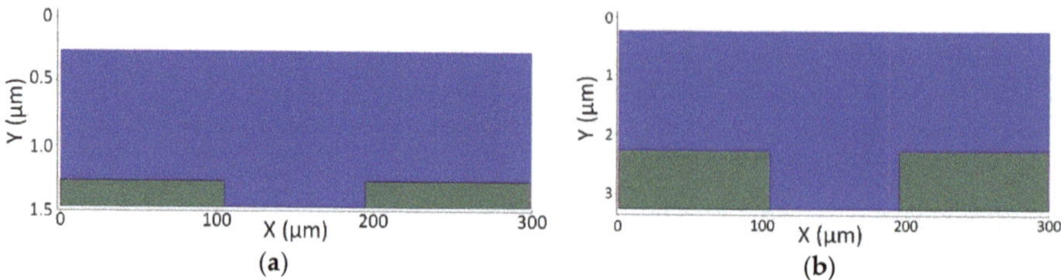

Figure 3. Device structure without ZnO/CdS layers. (a) CIGS and molybdenum thickness of 1 and 0.2 µm, respectively; (b) CIGS and molybdenum thickness of 2 and 1 µm, respectively.

Experimental procedure—We performed experiments on a real CIGS sample with monolithic interconnect to compare these results with the simulations of the models presented in Figures 2 and 3. The elimination of the ZnO/CdS layers was achieved through chemical etching with HCl (5% concentration) applied for 5 min. There is a previous publication related to the effect of the SCR on the conductivity of devices before and after etching presented in reference [18], although in this particular case, the experiment also involved dark annealing before the CdS layer was removed.

In our case, we measured the dark JV-curve of the CIGS sample at room temperature (25 °C) before and after etching without applying any other treatments to compare these results with the simulations. Our sample was produced in a pilot line with a co-evaporation process, a laser scribing for P1 and a mechanical patterning for P2 and P3, similar to reference [19]. Mechanical patterning of P2 and P3 is known to create fewer smooth trenches compared to laser ablation [20]. However, scribing P2 and P3 by nanosecond laser ablation is not recommended, as it might also damage the back contact [21]. According to reference [21], structuring of P2 and P3 scribes can be performed with picosecond laser ablation.

Alternative procedures for printing interconnect scribes in the device are presented in reference [22], in which P1, P2, and P3 are scribed with a femtosecond laser pulse after all the layers are deposited. P1 is filled then with a dielectric material and a metal on top of it to keep the electric continuity in the transparent contact oxide layer (TCO). On the other hand, P2 is filled with a metal to connect the TCO with the back contact of the next cell. Different materials for the filling were discussed in reference [22], including their impact on the electric behavior of the device. The method presented in reference [22] is expected to reduce the total dead area of one cell from 500 to 100 µm.

Previous research on thermally induced metastabilities provided a relation between the degradation of the electric characteristics in CIGS devices and their decrease in doping based on the results of capacitance profiling techniques; some of these can be found in references [23]. However, other works suggest that these accelerated ageing treatments induce other collateral effects such as enhancing the Schottky barrier between molybdenum and CIGS [24]. Thus, these treatments are not considered to study the transistor effect presented in this contribution.

3. Results

3.1. Results of the Simulations

Figure 2 presents the structure used for the first simulation, which is performed at 300 K. The results of this simulation are presented in Figure 4, with the JV-curve in the dark for varied N_A values in the CIGS bulk. A nonlinear decrease of reverse current density with decreasing N_A according to Figure 4b was observed. From the curves of Figure 4a it can also be deduced that the forward bias required to open the current path through the p–n junction decreases with N_A. This bias could be related to the built-in voltage (V_{bi}) of the device [16], which is the height of the barrier due to the SCR.

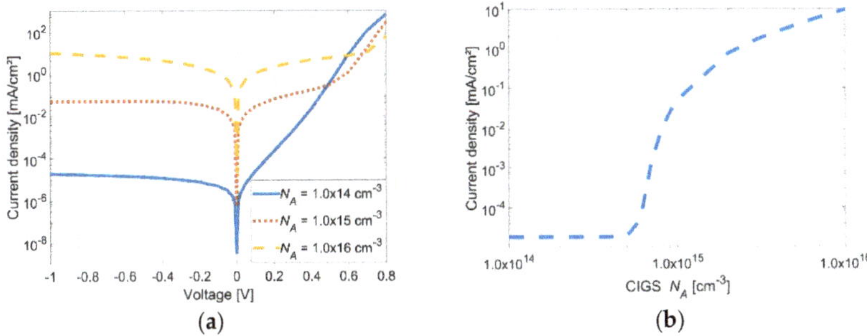

Figure 4. (a) Semilog *JV*-curve of the structure for varied values of N_A; (b) Current density at 1 V reverse bias for varied values of N_A.

Another interesting result that could be extracted from Figure 4a is the existence of a non-symmetric behavior between the forward and reverse bias regions for $N_A = 10^{14}$ cm^{-3}. This can be explained by the fact that for lower values of N_A, the non-linear diode behavior of the device starts to dominate over the linear one at lower positive biases compared to the case of higher CIGS dopings.

Figure 5 shows a 2D plot of the simulated current density. This result gives another hint regarding the transistor effect created by the SCR and the conductance through P1. If N_A in CIGS decreases, the SCR extends into P1 and reduces the available space in the channel for the current flowing through it. To the contrary, when N_A increases, the SCR shrinks to the junction, widening the channel.

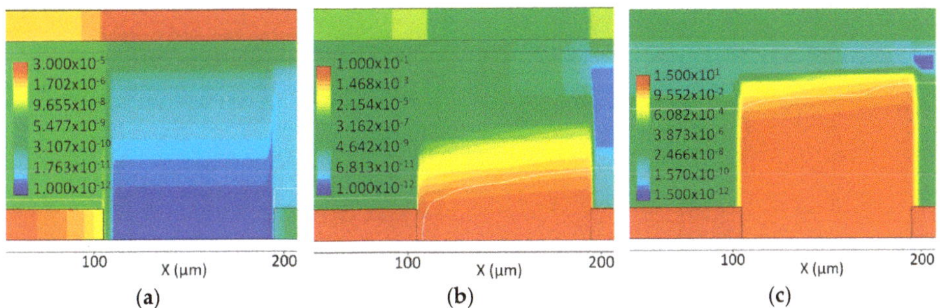

Figure 5. 2D representation of the current density flowing through the P1-shunt under a reverse bias of 1 V between anode and cathode for CIGS with: (a) $N_A = 10^{14}$ cm^{-3}; (b) CIGS $N_A = 10^{15}$ cm^{-3}; (c) $N_A = 10^{16}$ cm^{-3}. The color scale of each plot is in (mA/cm^2).

As mentioned above, V_{bi} is reduced when N_A decreases. It is possible to explain this behavior with the electric field inside the device across the p–n junction, [16]. In Figure 6a, a vertical cut of the vertical component of the electric field over the anode (see Figure 2a) for the extreme values of N_A is presented. The electric field in the case of $N_A = 10^{16}$ cm^{-3} has a higher value at the CdS/CIGS interface than in the case of $N_A = 10^{14}$ cm^{-3}. On the other hand, for $N_A = 10^{14}$ cm^{-3}, the electric field in the CIGS bulk is constant, indicating full carrier depletion down to the contact, while in the case of $N_A = 10^{16}$ cm^{-3}, the electric field in the CIGS bulk reduces linearly to 0, in support of the existence a quasi-neutral region with flat bands.

Further evidence of the effect of the SCR in the device is the absence of free carriers in the CIGS layer. These results belong to the simulation of the structure presented in Figure 2. In Figure 6b there is a representation of a horizontal cut between both contacts with the value of the free hole concentration. These curves correspond to values of $N_A = 10^{14}$ cm^{-3} and $N_A = 10^{16}$ cm^{-3}. With lower N_A, the concentration of free holes in P1 is also reduced, but not with a linear dependency on N_A.

In the case of $N_A = 10^{14}$ cm^{-3}, the concentration of free holes in P1 is 12 orders of magnitude lower than the value of N_A, while in the case of $N_A = 10^{16}$ cm^{-3}, the free hole density is in the same order of magnitude with the value of N_A. This absence of free holes means that the hole current density between both contacts will also be reduced in the same proportion when a reverse bias is applied.

The thickness of the CIGS layer is another key parameter for the behavior of the device, as mentioned previously. The SCR extends into the CIGS layer, and if the width of the channel is sufficiently reduced, it is possible to close the path of the current through the P1 region. In Figure 7a we present a comparison between the current density for a 2 and a 1-μm thick CIGS layer. A particular emphasis on the current density in the P1-shunt for the device with a 2-μm thick CIGS layer $N_A = 10^{15}$ cm^{-3} is presented in Figure 7b.

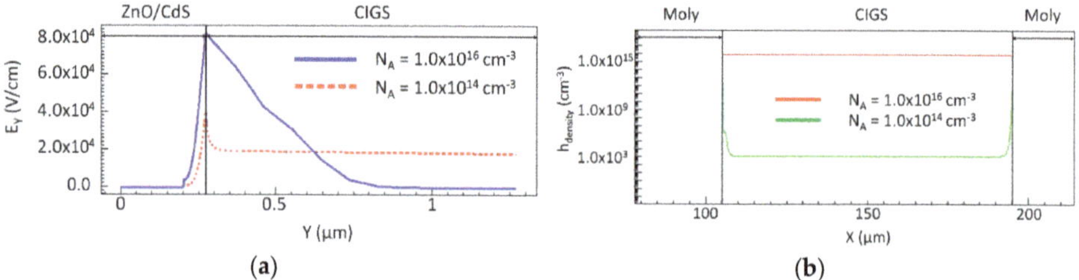

Figure 6. (**a**) Vertical cut (see Figure 2a) of the Y-component of the electric field (E_y) with $N_A = 10^{16}$ cm^{-3} and 10^{14} cm^{-3} under a bias between anode and cathode of 0 V; (**b**) Horizontal cut (see Figure 2a) of the free hole density between both molybdenum contacts for $N_A = 10^{16}$ and 10^{14} cm^{-3}.

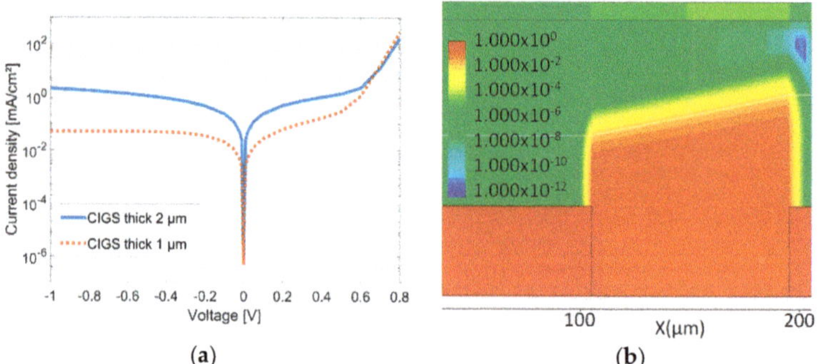

Figure 7. (**a**) JV-curve in the dark for the structures with 2 and 1 μm CIGS layer thickness, respectively and $N_A = 10^{15}$ cm^{-3}; (**b**) 2D plot of the current density flowing through P1 in the structure with 2 μm CIGS and 1 μm molybdenum thicknesses under reverse bias of 1 V, the color scale is in mA/cm^2.

If we compare the results presented in Figure 7b with those of Figure 5b, the SCR does not reach the bottom of P1 when the CIGS layer is thicker. In the structure with CIGS thickness of 1 μm, the SCR partially closes the path through P1, reducing the current density in reverse bias.

Figure 3a presents the structure of the second simulation with only CIGS/molybdenum layers, the latter divided by the P1 scribe. The temperature of the simulation was set at 300 K. A comparison between the JV-curves for different values of N_A in CIGS can be seen in Figure 8a, whereas a curve of the variation of the current in reverse bias for values of N_A between 10^{14} and 10^{16} cm^{-3} is shown in Figure 8b, where the current density increases linearly with the value of N_A.

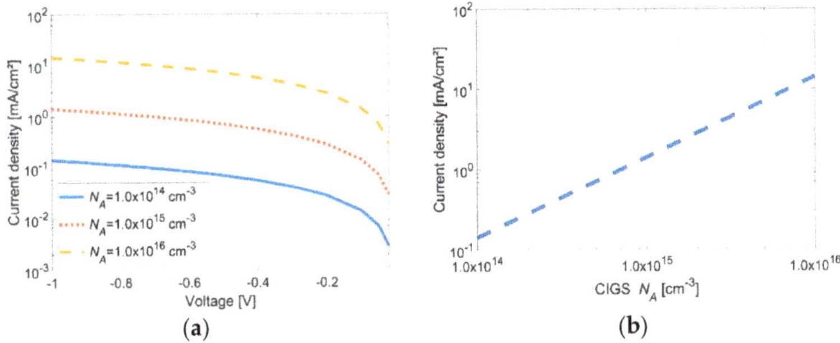

Figure 8. (**a**) Dark JV-curves of the device without ZnO/CdS for values of N_A between 10^{14} cm^{-3} to 10^{16} cm^{-3}; (**b**) Current density vs. N_A under a reverse bias of 1 V between both contacts.

3.2. Experimental Results

It is possible to provide some evidence regarding the effect of the SCR in the reverse current of real devices. As mentioned before, if the ZnO/CdS layers are removed from the device via HCl etching, it will be possible to measure the JV-curve without the impact of the SCR. The results of the JV-measurement at 25 °C are shown in Figure 9, compared to simulation results from the models presented in Figures 2b and 3b.

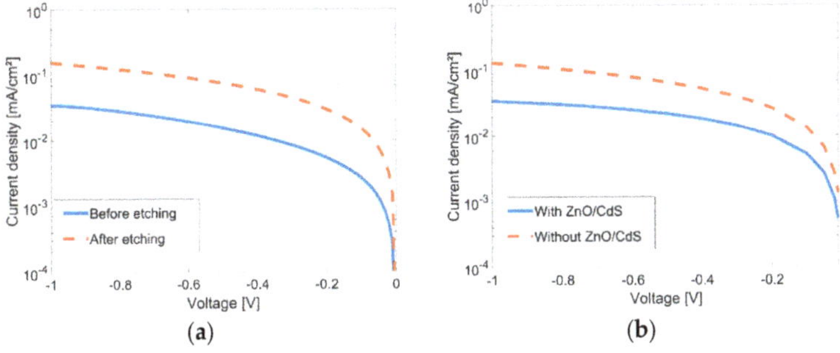

Figure 9. (**a**) JV-curve in reverse bias of the device before and after etching the ZnO/CdS layers; (**b**) Simulated JV-curves in reverse bias of the structures presented in Figure 2b (with ZnO/CdS layers) and Figure 3b (with no ZnO/CdS layer) with $N_A = 3 \times 10^{14}$ cm^{-3}. Electron and hole mobilities are reduced to 0.25 and 0.5 cm^2/V·s, respectively, to fit the experimental curves.

Both the experimental and simulated results show a similar behavior. When the ZnO/CdS layer was removed, the current density flowing between both contacts increased. As mentioned above, the simulation results relate this effect to the elimination of the SCR created due to the doping gradient between the n-type and p-type semiconductors. The semiconductor parameters of the CIGS sample used in the experiment are unknown, although it is possible to adjust the simulation parameters until a good compromise with experimental results is reached as indicated in Figure 9. A necessary reduction of hole and electron mobilities in order to fit experimental findings might be justified by the fact that the shunt path across P1 involves a lateral current transport across potential barriers due to grain boundaries, in agreement with [25]. The columnar nature of the grains leads to expect a higher number of grain boundaries in the horizontal direction (along the x-axis) compared to the vertical one (along the y-axis).

4. Discussion

The results of our work suggest a mechanism similar to the JFET simplified model presented in reference [9]. In our case, the equivalence of source and drain are both molybdenum contacts, and P1 is the channel. ZnO and CdS are part of the n-type gate material, which is directly connected to the drain. If a forward bias is applied, the current flows through the p–n junction instead of through the channel. Increasing or reducing the doping gradient between the p and n-type semiconductor layers affects the SCR in the device, and if this region is wide enough to close the channel, the P1 shunt conductance will be reduced.

This behavior is presented in Figure 4a, as the reverse current bias was significantly reduced when N_A in CIGS is decreased below 10^{15} cm^{-3}. Figure 4b provides further evidence of this transistor-like behavior, as the reverse bias current does not have a linear dependence with the applied bias. In fact, the reverse bias current is constant for values of N_A lower than 5×10^{14} cm^{-3}, as the SCR closes the path of the current through P1. If we increase this value, the current density in reverse bias increases exponentially as the SCR thickness is reduced and the path through P1 opens. After a certain value of N_A in CIGS, in this case 10^{15} cm^{-3}, the current density in reverse bias increases almost linearly. These results are in agreement with those presented in reference [11], in which decreasing the value of N_A below 10^{15} cm^{-3} is shown to have no impact on the shunt resistance of the device. It is interesting to mention that a gallium gradient, which is present in highly efficient CIGS devices with bandgap grading [26], can also play a role in the transistor-like behavior. A depth-dependent gallium distribution may induce a doping gradient [27], which to certain extent could affect the width of the SCR, and therefore the transistor effect might not be present. For sake of simplicity in the modelling, this effect has not been considered in our simulations, but could be regarded as a topic for further work.

The decrease of V_{bi} is a side-effect of manipulating the value of N_A. It should be considered carefully as it will also affect other parameters such as the open circuit voltage, the maximum power point, the fill factor and the efficiency. A good compromise between N_A and CIGS thickness is required to decrease the P1 shunt while reducing the impact on these parameters.

Removing the ZnO/CdS layer provides further evidence of the impact of the SCR in the P1 shunt and the reverse current density of the device. In this situation, the SCR is not present, and the reverse current density depends only on N_A; thus, the transistor effect is not present in the behavior of the device. In Figure 8b, the reverse current density increases proportional to N_A; in contrast with the results presented in Figure 4b. This increase in the reverse bias current is also presented in reference [18], in which the elimination of the SCR is considered to be the main reason behind this behavior.

This elimination of the transistor-like behavior is made plausible in the experimental results and the simulation of Figure 9. Etching away the ZnO/CdS layer increases the reverse current density. This increase may vary depending on the width of the SCR and the thickness of the CIGS layer. The results of our simulations are in agreement with experimental findings.

5. Conclusions

Based on 2D simulation results of CIGS cells with monolithic interconnects we proposed a mechanism that affects the P1 shunt, and consequently, the Ohmic behavior of the device. As has been discussed in this contribution, this mechanism has similarities with the behavior of a p-type JFET transistor device. According to this model, CIGS in P1 is the p-type channel in the device. As has been shown in the simulations, varying the thickness and the doping of CIGS, the width of the channel can be modulated, and therefore the current flowing in P1. Reducing N_A extends the SCR into the P1 interconnect. This means that the channel width is narrowed leading to a decreased shunt conductance. Experimental evidence was also provided to support the validity of the simulations. *JV*-measurements before and after etching the ZnO/CdS layer showed that when the n-type semiconductor was removed, the SCR and the transistor behavior were eliminated. The simulation results of a model with and without the ZnO/CdS layers support qualitatively the proposed mechanism.

Author Contributions: Conceptualization, R.V.L. and T.W.; Methology, R.V.L. and T.W.; Project Administration, T.W.; Supervision, D.F.M. and T.W.; Investigation, R.V.L.; Writing—Original Draft, R.V.L.; Writing—Review and Editing, T.W., T.L., D.M. and D.F.M.

Funding: This research was financed by the Federal Ministry for Economic Affairs and Energy of Germany under the proCIGS project (No. 0324070).

Acknowledgments: We would like to thank the Ulm University of Applied Sciences and IES-UPM (Instituto de Energía Solar, Universidad Politécnica de Madrid) for their support in this project.

Conflicts of Interest: The authors declare no conflict of interest.

References

1. Jackson, P.; Wuerz, R.; Hariskos, D.; Lotter, E.; Witte, W.; Powalla, M. Effects of heavy alkali elements in Cu(In,Ga)Se$_2$ with efficiencies up to 22.6%. *Phys. Status Solidi RRL Rapid Res. Lett.* **2016**, *10*, 583–586. [CrossRef]
2. Green, M.A.; Hishikawa, Y.; Dunlop, E.D.; Levi, D.H.; Hohl-Ebinger, J.; Ho-Bailie, W.A. Solar cell efficiency tables (version 52). *Prog. Photovolt. Res. Lett.* **2018**, 427–436. [CrossRef]
3. Green, M.A.; Emery, K.; Hishikawa, Y.; Warta, W.; Dunlop, E.D. Solar cell efficiency tables (version 48). *Prog. Photovolt. Res. Appl.* **2016**, *24*, 905–913. [CrossRef]
4. Yang, S.; Lin, K.-M.; Lee, W.-C.; Lo, W.-S.; Chen, C.-H.; Wu, J.-L.; Chiang, C.-Y.; Wu, C.-H.; Sun, Y.-L.; Lo, H.; et al. Achievement of 16.5% total area efficiency on 1.9 m^2 CIGS modules in TSMC solar production line. In Proceedings of the IEEE 42nd Photovoltaic Specialist Conference (PVSC), New Orleans, LA, USA, 14–19 June 2015. [CrossRef]
5. Contreras, M.A.; Egaas, B.; Ramanathan, K.; Hiltner, J.; Schwartzlander, A.; Hasoon, F.; Noufi, R. Progress towards 20% efficiency in Cu(In,Ga)Se$_2$ polycristalline thin-film solar cells. *Prog. Photovolt. Res. Appl.* **1999**, *7*, 311–316. [CrossRef]
6. Scheer, R.; Schock, H.-W. *Chalcogenide Photovoltaics: Physics, Technologies and Thin Film Devices*; Wiley-VCH Verlag GmbH & Co.: Weinheim, Germany, 2011; pp. 9–127.
7. Portmans, J.; Arkhipov, V. *Thin Film Solar Cells: Fabrication, Characterization and Applications*; John Wiley & Sons: Chichester, UK, 2006; pp. 237–275.
8. Brecl, K.; Topič, M.; Smole, F. A detailed study of monolithic contacts and electrical losses in a large-area thin-film module. *Prog. Photovolt. Res. Appl.* **2015**, *13*, 297–310. [CrossRef]
9. Yoon, J.H.; Park, J.K.; Kim, W.M.; Lee, J.; Pak, H.; Jeong, J.H. Characterization of efficiency-limiting resistance losses in monolythically integrated Cu(In,Ga)Se$_2$ solar modules. *Sci. Rep.* **2015**, *5*, 7690. [CrossRef] [PubMed]
10. Neamen, D.A. *Semiconductor Physics and Devices: Basic Principles*, 4th ed.; McGraw-Hill: New York, NY, USA, 2012; pp. 571–611.
11. Schubbert, C.; Eraerds, P.; Richter, M.; Parisi, P.; Riedel, I.; Dalibor, T.; Palm, J. Performance ratio study based on a device simulation of a 2D monolithic interconnected Cu(In,Ga)(Se,S)$_2$ solar cell. *Sol. Energy Mater. Sol. Cells* **2016**, *157*, 146–153. [CrossRef]
12. Sozzi, G.; Pignoloni, D.; Menozzi, R.; Pianezzi, F.; Reinhard, P.; Bissig, B.; Buecheler, S.; Tiwari, A.N. Designing CIGS solar cells with front-side point contacts. In Proceedings of the IEEE 42nd Photovoltaic Specialist Conference (PVSC), New Orleans, LA, USA, 14–19 June 2015. [CrossRef]
13. Park, J.; Shin, M. Numerical optimization of gradient bandgap structure for CIGS solar cells with ZnS buffer layer using Technology Computer-Aided Design simulation. *Energies* **2018**, *11*, 1785. [CrossRef]
14. Movla, H. Optimization of the CIGS based thin-film solar cells: Numerical simulations and analysis. *Optik* **2014**, *125*, 67–70. [CrossRef]
15. Synopsys. Available online: https://www.synopsys.com/silicon/tcad.html (accessed on 3 December 2018).
16. Sze, S.M.; Ng, K.K. *Physics of Semiconductor Devices*, 3rd ed.; John Wiley & Sons: Hoboken, NJ, USA, 2007; pp. 79–124.
17. Li, S.; Fu, Y. *3D TCAD Simulation for Semiconductor Processes, Devices and Optoelectronics*; Springer: New York, NY, USA, 2012; pp. 41–80.
18. Lavrenko, T.; Vidal Lorbada, R.; Muecke, D.; Walter, T.; Plesz, B.; Schaeffler, R. Towards an improved understanding of CIGS thin film solar cells. In Proceedings of the 33rd European Photovoltaic Solar Energy Conference and Exhibition, Amsterdam, The Netherlands, 25–29 September 2017. [CrossRef]

19. Powalla, M.; Cemernjak, M.; Eberhardt, J.; Kessler, F.; Kniese, R.; Mohring, H.D.; Dimmler, B. Large-area CIGS modules: Pilot line production and new developments. *Sol. Energy Mater. Sol. Cells* **2006**, *90*, 3158–3164. [CrossRef]

20. Westin, P.O.; Zimmermann, U.; Edoff, M. Laser patterning of P2 interconnect via in thin-film CIGS PV modules. *Sol. Energy Mater. Sol. Cells* **2008**, *92*, 1230–1235. [CrossRef]

21. Heise, G.; Boerner, A.; Dickmann, M.; Englmaier, M.; Heiss, A.; Kemnitzer, A.; Konrad, J.; Moser, R.; Palm, J.; Vogt, H.; et al. Demonstration of the monolithic interconnection on CIS solar cells by picosecond laser structuring on 30 by 30 cm^2 modules. *Prog. Photovolt. Res. Appl.* **2015**, *23*, 1291–1304. [CrossRef]

22. Fields, J.D.; Dabney, M.S.; Bollinger, V.P.; van Hest, M.F. Printed monolithic interconnects for photovoltaic applications. In Proceedings of the IEEE 40th Photovoltaic Specialist Conference (PVSC), Denver, CO, USA, 8–13 June 2014. [CrossRef]

23. Mack, P.; Ott, T.; Walter, T.; Hariskos, D. Optimization of reliability and metastability of CIGS solar cell parameters. In Proceedings of the 25th European Photovoltaic Solar Energy Conference and Exhibition, Valencia, Spain, 6–10 September 2010. [CrossRef]

24. Ott, T.; Lavrenko, T.; Walter, T.; Schaeffler, R.; Fetch, H.-J. On the importance of the back contact for Cu(In,Ga)Se$_2$ thin film solar cells. In Proceedings of the 29th European Photovoltaic Solar Energy Conference and Exhibition, Amsterdam, The Netherlands, 22–26 September 2014. [CrossRef]

25. Abou-Ras, D.; Schmidt, S.S.; Schaefer, N.; Kavalakkatt, J.; Rissom, T.; Unold, T.; Mainz, R.; Weber, A.; Kirchartz, T.; Sanli, E.S.; et al. Compositional and electrical properties of line and planar defects in Cu(In,Ga)Se$_2$ thin films for solar cells—A review. *Phys. Status Solidi Rapid Res. Lett.* **2016**, *10*, 363–375. [CrossRef]

26. Dullweber, T.; Hanna, G.; Shams-Kohali, W.; Schwartzlander, A.; Contreras, M.A.; Noufi, R.; Schock, H.W. Study of the effect of gallium grading in Cu(In,Ga)Se$_2$ with efficiencies up to 22.6%. *Thin Solid Films* **2000**, *361*, 478–481. [CrossRef]

27. Frisk, C.; Platzer-Bjoerkman, C.; Olsson, J.; Szaniawski, P.; Waetjen, J.T.; Fjaellstroem, V.; Salomé, P.; Edoff, M. Optimizing Ga-profiles for highly efficient Cu(In,Ga)Se$_2$ thin-films in simple and complex defect models. *J. Phys. D Appl. Phys.* **2014**, *47*, 485104. [CrossRef]

 *coatings*

Article

Enhancement for Potential-Induced Degradation Resistance of Crystalline Silicon Solar Cells via Anti-Reflection Coating by Industrial PECVD Methods

Tsung-Cheng Chen [1,2], Ting-Wei Kuo [2], Yu-Ling Lin [2], Chen-Hao Ku [2], Zu-Po Yang [1,*] and Ing-Song Yu [3,*]

1   Institute of Photonic System, National Chiao Tung University, Tainan 71150, Taiwan; bob.chen@e-tonsolar.com
2   E-ton Solar Tech Co., Ltd., Tainan City 709, Taiwan; E0709003@e-tonsolar.com (T.-W.K.); johnny.lin@e-tonsolar.com (Y.-L.L.); chenhao@e-tonsolar.com (C.-H.K.)
3   Department of Materials Science and Engineering, National Dong Hwa University, Hualien 97401, Taiwan
*   Correspondence: zupoyang@nctu.edu.tw (Z.-P.Y.); isyu@gms.ndhu.edu.tw (I.-S.Y.); Tel.: +886-6-303-2121 (Z.-P.Y.); +886-3-890-3219 (I.-S.Y.)

Received: 7 October 2018; Accepted: 21 November 2018; Published: 22 November 2018

Abstract: The issue of potential-induced degradation (PID) has gained more concerns due to causing the catastrophic failures in photovoltaic (PV) modules. One of the approaches to diminish PID is to modify the anti-reflection coating (ARC) layer upon the front surface of crystalline silicon solar cells. Here, we focus on the modification of ARC films to realize PID-free step-by-step through three delicate experiments. Firstly, the ARC films deposited by direct plasma enhanced chemical vapor deposition (PECVD) and by indirect PECVD were investigated. The results showed that the efficiency degradation of solar cells by indirect PECVD method is up to −33.82%, which is out of the IEC 62804 standard and is significantly more severe than by the direct PECVD method (−0.82%). Next, the performance of PID-resist for the solar cell via indirect PECVD was improved significantly (PID reduced from −31.82% to −2.79%) by a pre-oxidation step, which not only meets the standard but also has higher throughput than direct PECVD. Lastly, we applied a novel PECVD technology, called the pulsed-plasma (PP) PECVD method, to deal with the PID issue. The results of the HF-etching rate test and FTIR measurement indicated the films deposited by PP PECVD have higher potential against PID in consideration of less oxygen content in this film. That demonstrated the film properties were changed by applied a new control of freedom, i.e., PP method. In addition, the 96 h PID result of the integrated PP method was only −2.07%, which was comparable to that of the integrated traditional CP method. In summary, we proposed three effective or potential approaches to eliminate the PID issue, and all approaches satisfied the IEC 62804 standard of less than 5% power loss in PV modules.

Keywords: anti-reflection coating; potential-induced degradation; solar cell; plasma enhanced chemical vapor deposition

1. Introduction

Photovoltaic (PV) has been recognized as the most competitive renewable energy among various renewable technologies according to its diverse applications and is a candidate of next generation large-volume power plants. For instance, Águas et al. [1] applied the PV for building-integrated photovoltaics (BIPV) application; Águas et al. [2] also fabricated thin film solar cells on cellulose paper for flexible, wearable application; and most of all was for large-scale power plants. According to the statistic reports of International Energy Agency (IEA), it pointed out that the cumulative installed capacity of solar installations was almost 303 GW at the end of 2016 [3]. Based on the demand for long-term and safe operation, the reliabilities of PV modules gain more attention. In Figure 1, we depicted the components of crystalline silicon module under negative bias. Solar cells were held in between the front glass cover and the rear backsheet. The glass was used for light transmission and module protection; the backsheet was implemented for the reflection of incident lights shined at the gap between individual solar cells and for the reflection of residual lights that penetrate the solar cells. In general, the lifetime of PV modules was designed over 25 years. However, several kinds of issues resulted in irreversible degradation of PV modules, in which one of the most catastrophic failures was noted as potential-induced degradation (PID).

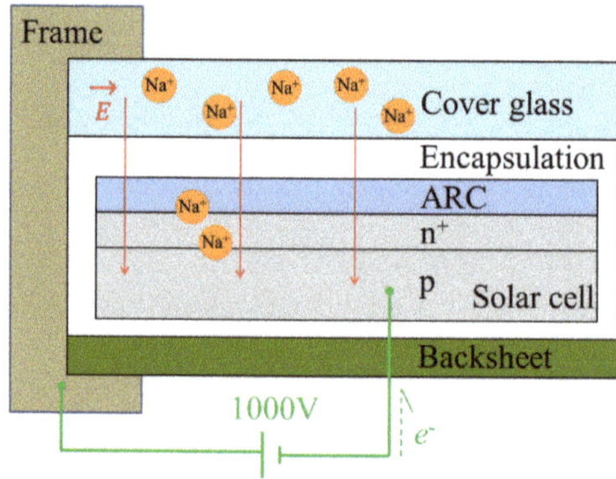

Figure 1. The structure of a crystalline Si PV module was introduced. In this diagram, the solar cells are negative-biased and the shunting model via sodium ions (Na$^+$) is depicted.

PID was firstly reported for both crystalline silicon modules and amorphous silicon thin film modules by the Jet Propulsion Laboratory in 1985, but it was observed by SunPower Corp. until 2005 and coined by Pingel et al. in 2010 [4,5]. PID is ascribed to a high electric potential difference across a PV module inducing a leakage current flowing through the solar cells, which results in reversible or irreversible power conversion efficiency degradation. The schematic diagram is shown in Figure 1. Currently, 1000 V is applied in a PV power system to minimize the electric power transportation loss by serially connected PV modules. Nevertheless, the PID of PV modules is significantly enhanced under a high bias. For p-type Si-based solar cells, the most remarkable type of PID mechanism is "potential induced shunting" (PID-s). PID-s describes a degradation ascribed to both sodium ions and solar cells under a negative-biased condition. Sodium ions (Na$^+$), coming from soda-lime glass, are drifted through the anti-refractive coating (ARC) layer, and then accumulate at the ARC/Si interface or in the stacking faults. Finally, the shunting paths are formed, and then the p-n junction is damaged.

Several promising strategies to approach PID-resistant PV systems have been proposed for cell-, module-, or system-levels [6,7]. On the cell-level, modifying the ARC layer by elevating its refractive index (RI) has well been demonstrated to alleviate the PID-s issue [5,7–10]. In the factory of c-Si solar cells, various plasma enhanced chemical vapor deposition (PECVD) methods provide the ARC layers for solar cells, including direct, remoter, continuous-plasma (CP) and pulsed-plasma (PP) types. From the literature, it was reported that increasing the RI of the ARC layer improves its electric conductivity which can prevent positive charges accumulating in the surface and Na$^+$ drifted by electric field. However, if the RI of the ARC layer deviates from the theoretically optimal value (i.e., the square-root of the RI of silicon), the light trapping performance of ARC degrades. Hence, there is a trade-off between PID-resistant ability and power conversion efficiency (PCE). In addition, Mishina et al. [11] pointed out that the film quality of the ARC has a great impact on PID-resistant ability. They showed that the ARC films deposited by the novel direct PECVD performed better PID-resistance than the one deposited by remoter PECVD, even without scarifying the PCE (i.e., keeping the conventional RI value). However, direct PECVD, in comparison with remoter PECVD, has relatively low throughput and gains additional running costs in commercialized applications. Beside the modification of ARC films, a thin oxide film between the ARC and Si substrate has been proven as another efficient solution for PID-resistant solar cells [7,12,13]. Furthermore, Nagel et al. demonstrated that the cells with this thin thermal oxide layer gained a 0.2% increase in PCE, which is ascribed to the improvement of surface passivation in the front surface [12].

Recently, a novel PECVD method, PP modulated PECVD, is commonly used for modifying SiN_x films by controlling the ion bombardment and film nucleation process [14–17]. Contrary to continuous-plasma PECVD, the manipulation of pulsed-plasma PECVD is by using a frequency modulated glow discharge during deposition, and then modifying the film qualities. After ionization, the lifetimes between free radicals remain different, and then the clusters during the film synthesization have been designated by controlling the duty cycle and periods. The duty cycle is the ratio of conducting glow discharge to the periods, as shown in the schematic diagram in Figure 2. In the scenario of low frequency modulated radio frequency (RF) discharge (40 Hz) in Watanabe et al.'s study [18], lifetimes of SiH_n (n = 0–2) radicals were shorter than 10 ms (these were estimated to be below 3 ms); whereas the lifetime of SiH_3 was longer than 20 ms. Regarding the achievement of film quality modification by PP PECVD, Watanabe et al. also pointed out the film mainly composed of SiH_3 shows fewer dangling bonds and reduced powder particle in size during deposition [18]. In addition, Viera et al. reported a high purity and controllable nanostructure achieved by the PP method [19], and Byungwhan Kim et al. pointed out that the films made by the PP method present denser and smoother surface roughness [20].

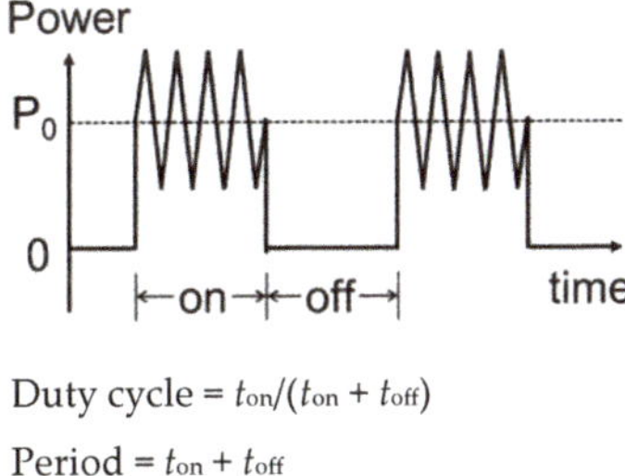

Duty cycle = $t_{on}/(t_{on} + t_{off})$

Period = $t_{on} + t_{off}$

Figure 2. Time modulated RF source and the definition of duty cycle and of period.

The PID issue has been addressed for years. Current PV power plants have enrolled different kinds of cutting edge technologies to solve uprising PID concern. For instance, other types of solar cells (PERC and HIT) with inherent high voltage output are becoming mainstream, and 1500 V power system are replacing traditional 1000 V power systems [10,21]. In consideration of even stricter PID-resistant demands, improved or novel PID-resistant solutions are necessary in the future. In this work, we firstly proposed a PP PECVD technology to deal with the PID issue. Besides, the comparisons of the ARC films deposited by direct and indirect PECVD, and the enhancement of the PID-resistant ability by the additional interface oxide layer were investigated.

2. Materials and Methods

In this work, three closely related experiments were designed to approach PID-free c-Si solar cells. The cell structure is depicted in Figure 1. First, we compared PID-resistant abilities between different kinds of commercially available PECVD, namely direct PECVD (Shimadzu Corporation, Kyoto, Japan) and indirect PECVD (OTB SOLAR DEPX 2400). For the direct PECVD system, the electrodes are placed across the specimens, and the reactant gases are ionized nearby the specimens. Besides, the electric field across the electrodes can boost the moment of ions and thus these energetic ions bombard the deposited film during deposition. Contrary to direct PECVD, the gaseous ions are generated separately from the substrates and sent to target locations through an inlet for the indirect PECVD. Secondly, in consideration of the relatively higher throughput of the OTB system, we introduced a thin oxide layer at the interface of the ARC film and Si to solve the PID issue. For these two experiments, we used 6-inch multi-crystalline silicon (mc-Si) solar cells with traditional Al back surface field (BSF), and these specimens were tested by the standard PID testing procedure. Moreover, we investigated the SiN_x films deposited by PP and CP PECVD (Otb Solar Depx 2400 by Roth & Rau, Hohenstein-Ernstthal, Germany). In common, there are two approaches to modulate the glow discharge for realizing the PP method by either controlled inlet gas or controlled bias. The latter one was employed in this work. The deposition parameters of RF frequency, modulated frequency, and duty cycle are 13.56 MHz, 80 Hz and 50%, respectively. For the PID test, we replaced the conventional mc-Si solar cells by the PERC-type mono-crystalline silicon (c-Si) solar cells which had a thin oxide layer located at the interface of the ARC and Si as well. For the characterizations of these films, we also conducted a dilute HF etching test. The dilute HF-etching rate has been pointed out as a good indicator of the film quality against PID [11]. A total of eight specimens with an RI value of 2.06 deposited by the CP and PP method were dipped in 3.85 vol % HF solution at room temperature. Fourier transform infrared (FTIR) spectra were conducted for the chemical composition of SiN_x films.

Based on the instruction of standard reliability test of PID, IEC-62804-1 [22], we followed and even tested a more severe testing condition to clarify the significance of experimental variables. All specimens in this experiment were carried out by the same PID testing procedures as following:

(1) Four solar cells were laminated in one module but disconnected from each other as shown in Figure 3.

(2) The electric performances and electroluminescence (EL) images of the PV modules were recorded before PID testing.

(3) Aluminum foils with deionized water wetted were attached upon the front glasses of modules, and then connected with the positive terminal of a high-voltage power source.

(4) On the other hand, the negative terminal of this high-voltage power source was connected to the shunted cathodes and anodes of solar cells.

(5) PV modules were processed in a climate chamber for 48/96 h, at 85 °C, with 85% relative humidity, and −1000 V bias.

(6) After PID treatment, the measurements referred to in Step 2 were repeated.

Figure 3. The photo of the (PID) test module with four cells packaged separately.

The equation for characterizing PID is listed as Equation (1):

$$PID\ (\%) = \frac{\left(P_{mpp,AF} - P_{mpp,BF}\right)}{P_{mpp,BF}} \tag{1}$$

where $P_{mpp,AF}$ is the maximum output power under standard testing conditions after PID treatment, and $P_{mpp,BF}$ is the one before PID treatment. Under the IEC 62804 standard, modules with less than 5% power loss and without induced major defects could be recognized as PID resistance [23,24].

3. Results and Discussion

3.1. PID-Resistant Approach via Film-Quality Modification

The PID-resistance of two film deposition methods was compared in this experiment, i.e., direct and indirect PECVD. The detail comparison of our equipment of these two approaches and the corresponding results are listed in Table 1. After performing the PID test, the results for mc-Si solar cells with SiN_x films deposited by the direct and indirect PECVD are presented in Table 2, showing the EL images before and after the PID test and power conversion efficiency loss. For the EL images, before PID treatment, both kinds of cells perform uniform illumination under forward bias. The EL image of the cells via indirect PECVD is relatively less uniform than via direct PECVD, which is ascribed to the inherent less gas inlets in indirect PECVD equipment. After the PID test, the apparent dark areas (marked by a red oval) were observed for the EL image of the cells via indirect PECVD, but barely seen for the direct PECVD. It indicates the more severe power conversion efficiency loss for the cell via indirect PECVD. The power conversion efficiency loss of the cell via the indirect PECVD method is up to -33.82%, significantly more severe than the one via the direct PECVD method (-0.82%). The highlighted areas in the red circles in Table 2 are located between two bus bars. This phenomenon can be interpreted by the PID-shunt model [7], where Na^+ ions are firstly accumulated on the surface of the ARC driven by the applied negative bias, and are then drifted through the ARC layer, and finally shunt the p-n junction dramatically.

Table 1. The comparison of direct and indirect PECVD and the controlled film thickness and refractive index of specimens.

Features	Direct PECVD	Indirect PECVD
RF system	Direct plasma	Remote plasma
Pressure control	Pressure control (APC valve)	Pump speed control (pump frequency)
Pressure	67 Pa	23–25 Pa
Deposition rate	Low speed	High speed
Process time	120 s	18 s
Temperature control	Set temperature	Set heater power
Process temperature	450 °C	450 °C
Gas flow ratio (SiH_4:NH_3)	1:3	1:3
Film thickness	85 nm	85 nm
Refractive index	2.10	2.10

Comparing the two deposition methods, the SiN_x films deposited by direct PECVD are denser, more compact, and more conformal, but had more surface damages due to ions bombardment. Therefore, we deduce that the denser SiN_x film made by direct PECVD can resist Na^+ penetration due to less pin holes and smaller clusters during deposition. However, its lower productivity is a concern from the point-of-view of industrial applications [11,16,25,26].

Table 2. The PID testing results and EL images for the solar cells with SiN_x films deposited by the direct and indirect PECVD.

Deposition Methods	Before PID	After PID	PID Results
ARC by direct PECVD			−0.86%
ARC by indirect PECVD			−33.82%

3.2. PID-Resistant Approach via Pre-Oxidation Treatment

Although Si solar cells via indirect PECVD is associated with severe PID concern, we note the advantages of ultrahigh deposition rate (5 nm/s), which has high throughput for mass production [26]. A pre-oxidation step before ARC films by indirect PECVD was introduced to improve PID-resistance ability. Therefore, a thin oxide layer (SiO_2 or SiO_xN_y) was formed between ARC and Si substrate by using a furnace with an atmosphere of O_2 and N_2 and with the setting temperature of 800 °C. The thickness of this oxide layer was around 2 to 5 nm confirmed by ellipsometer. In this section, the specimens with and without this interface oxide layer were produced. The PID testing results are presented in Table 3. The PID effect was dramatically improved and was down to −2.79% by inserting an oxide layer, which well fulfills the IEC 62804 standard of power loss less than −5%. The root cause about the role of the oxide layer was explained by Naumann et al. [27], and the extensive research about the role of oxide layer in the recovery process of PID was reported by Lausch et al. [28]. Volker Naumann pointed out two facts by investigating the cross section of the solar cell after PID treatment by TEM and EDX mapping. (1) Na was accumulated at the c-Si stacking fault resulting in the creation of a shunting path. (2) The results of O mapping identified the position of the interface oxide layer, and, moreover, the Na was also "trapped" and scatteringly distributed in the oxide layer. Lausch et al. further proved that this oxide layer may assist the Na to diffuse out from c-Si stacking fault when performing a PID recover experiment. In addition, the oxide layer served as a barrier against Na drift into the silicon base. In short, the PID-resistant ability can be improved by a factor of 10 in this experiment by introducing a thin oxide layer between Si and SiN_x films in the case of indirect PECVD which can fit the high-throughput requirement of the fabrication of Si solar cells.

Table 3. The PID testing results and EL images for the solar cells with or without an interface oxide layer between the ARC and Si substrate on indirect PECVD.

Approaches	Before PID	After PID	PID Results
With interface oxide layer			−2.79%
Without interface oxide layer			−31.82%

3.3. A Novel PID-Resistant Approach via SiN$_x$ Film Deposited by PP PECVD

Recently, PP PECVD is becoming the mainstream technique for the fabrication of c-Si solar cells. This novel method has some remarkable advantages, including higher throughput, no additional process step, being easier to control the film quality, and so on. Hence, we firstly proposed a novel study to improve the PID-resistance of SiN$_x$ films by PP PECVD. In addition, regarding higher PID-resistant demands for modern solar cells, we evaluated the PID-resistance of 6-inch c-Si PERC-type solar cells. For the characterization of these films, we conducted a dilute HF etching test, FTIR measurement, and PID test for SiN$_x$ films deposited by CP and PP methods.

The etched film thickness for different etching duration is presented in Figure 4. The film thicknesses of origin, and after 3, 6, and 9 min etched, were measured by ellipsometer, and the discrepancies were calculated. The HF-etching rates were deduced by linear fitting. The HF-etching rate of SiN$_x$ film deposited by PP PECVD is 2.5 nm/min, 2.6 times lower than the one by CP PECVD. Mishina et al. [11] pointed out that the mechanism of HF etching is associated with the SiN$_x$ dissociations via oxygen assistance, i.e., higher oxygen density in the SiN$_x$ film leads to a larger HF-etching rate. In addition, the SiN$_x$ film with lower HF etching rate (lower oxygen density) is related to high PID-resistance [11]. The root cause is that the oxygen contained in SiN$_x$ may be in the form of SiN$_2$O or SiO$_2$, and subsequently, the SiO$_2$ dissolved in the molten salt (Na$^+$) will form a liquid sodium silicate (Na$_2$O·(SiO$_2$)) [11,29,30]. The oxygen-assisted corrosion of the SiN$_x$ film in molten salt would create a shunting path and then worsens the original PID-resistance of the ARC film. Accordingly, since our ARC films made by the PP method contain less oxygen atoms than by the CP method, we derived the SiN$_x$ film bade by the PP method to perform better PID-resistance.

FTIR measurement was employed to characterize the chemical composition discrepancies between the films deposited by CP and by PP methods, and the results are shown in Figure 5. The absorbance at 850 cm^{-1} was designated as the asymmetric stretching vibration mode of Si–N bonding [31]; the absorbance at 2160 cm^{-1} was designated as the in phase stretching vibration mode of Si-rich Si–H bonds [32]. Compared to the peak of Si–N stretching mode of the SiN$_x$ film deposited by the PP method located at 831.2 cm^{-1}, the one deposited by the CP method is at 881 cm^{-1}. Shinichi Kobayashi reported that the large blue shift of the Si–N stretching mode is ascribed to oxidation because of the net change of electronegativity around the Si–N bonds due to the increase in the Si–O bond density [33]. The blue-shifted peak of Si–N mode indicates that the CP method-deposited SiN$_x$ film contains inherently higher oxygen, which agrees with the result of the dilute HF etching test. These results also suggest that the PP method can diminish the oxygen content in SiN$_x$ as well as can increase the PID-resistant ability. The absorbance peak at 3350 cm^{-1} is designated as N–H bonds [34], and these N–H bonds present almost identical for the films deposited both PP and CP methods. In addition, Si–N and Si–H peak magnitudes of the PP method-deposited SiN$_x$ film are significantly less than that of the CP method-deposited SiN$_x$ film. These results indicate that our PP method-deposited SiN$_x$ film is either relatively thinner or has a smaller RI value (or less Si content) than that of the CP method-deposited one. Ideally, to compare the PID-resistant ability of SiN$_x$ films deposited by CP and PP methods, it should have the same thickness and Si content. Therefore, these observed either thinner or lower Si content films might lead us under evaluate the anti-PID ability of the PP method SiN$_x$ film because Na$^+$ can easier penetrate into thinner and accumulate on lower conductivity SiN$_x$ films. Nevertheless, we succeeded in the modification of the SiN$_x$ property of film thickness, silicon content, or oxygen content by introducing the PP method as an additional control of freedom.

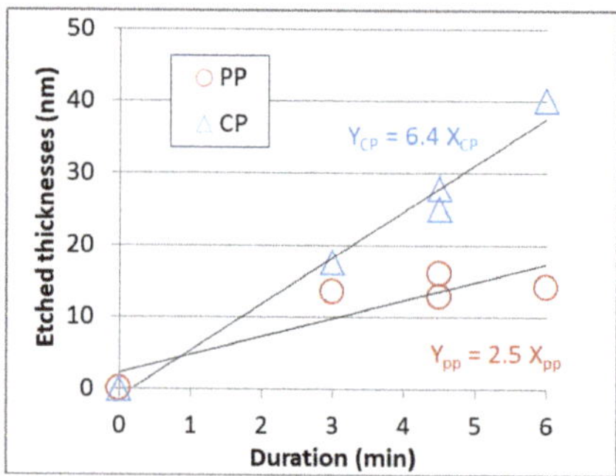

Figure 4. Etched thicknesses of SiN$_x$ films deposited by PP and CP PECVD in 3.85 vol % HF solution. The film thicknesses were measured by ellipsometer. Each specimen was recorded at five locations and the variations may come from the non-uniform etching rate in whole wafers and from the measuring uncertainties.

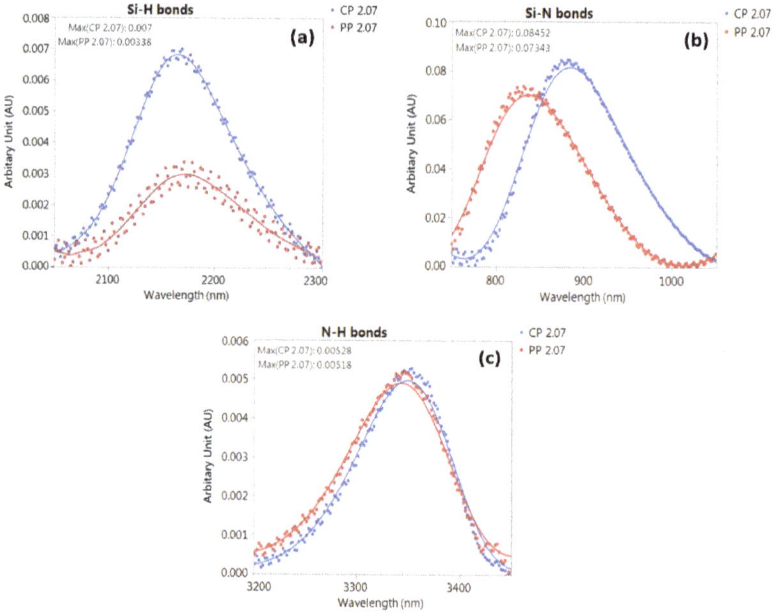

Figure 5. IR absorbance of (**a**) Si–H, (**b**) Si–N, and (**c**) N–H bonds by Fourier transform infrared (FTIR) characterizations for the SiN$_x$ films deposited by the CP and PP methods. The smooth curves are provided as a guide to the eye for indicating the absorbance peaks.

Finally, not only the 48- but also 96-h PID treatments for c-Si PERC-type solar cells, in which there were four and two pieces respectively, were performed and are shown in Figure 6. Simultaneously, EL images are shown in Table 4. PID results revealed that both groups of samples fit the requirement of IEC 62804 specification, i.e., power conversion efficiency loss of the cells after PID test is under 5%. It was noticeable that the PP group for 48-h testing had shown the largest error bar. This is because we retained all data and one of them seemed to be an outlier (−1.07%, −1.09%, −1.23%, −3.30%). As the duration of PID test increases, the power conversion efficiency loss gets higher.

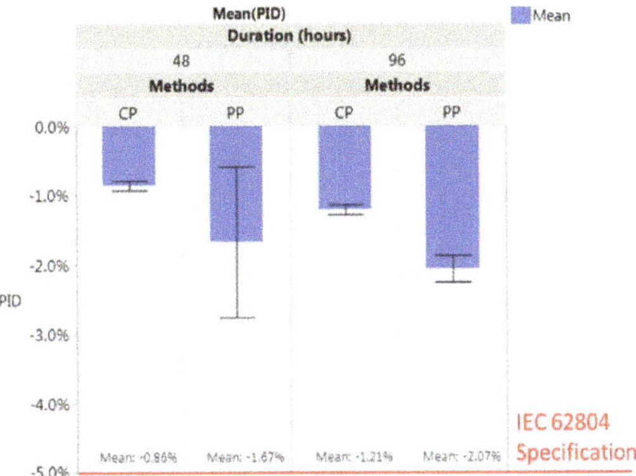

Figure 6. Distributions of 48/96 h PID testing results for the solar cells with CP and PP deposited anti-reflection coating. The red line represents the IEC 62804 specification (−5.0%).

Table 4. The 48/96 h PID testing results and EL images for the solar cells with CP and PP deposited anti-reflection coating.

Approaches	Before PID	After 48 h	After 96 h
CP method		PID −0.8%	PID −1.2%
PP method		PID −1.2%	PID −2.2%

The PID-resistant ability of the PP group was slightly lower than the one of the CP group. Since the oxide layer between the ARC film and Si plays an important role in the resistance of PID, the weighting of the PID-resistance ability via the PP modulated PECVD is reduced. Another reason is that the realistic deposition condition of the PP modulated PECVD is slightly away from our expectation. We thought that we had controlled the same RI value (2.07) for both groups and controlled comparable film thicknesses (of 78.6 nm (CP) and 77.9 nm (PP)) by single wavelength ellipsometer at 632.8 nm). However, by taking an empirical relation of refractive index with the stoichiometry from Dauwe [35] or complemented to low-pressure PECVD from Lelièvre et al. as below [36]:

$$n = 1.22 + 0.61 \frac{[\text{Si}]}{[\text{N}]} \tag{2}$$

The empirical RI can be estimated through chemical stoichiometry of Si and N. Hence, according to our FTIR results, the film deposited by PP PECVD contains a lower Si concentration, and thus has a lower RI value than the CP method-deposited SiN$_x$ film. Therefore, the extracted RI value by single wavelength ellipsometer might not be that accurate, and influences this delicate experiment at the beginning.

To the best of our knowledge, we may be the first team studying PID-resistant ability through the pulsed-plasma indirect PECVD method. This additional control of freedom was realized in adjusting film properties, which was confirmed by the mentioned dilute HF etching test and FTIR measurement. The PID results of the PP method are qualified for the generalized criteria of less than -5%, and has potentially achieved lower power loss if fine-tuning the SiN_x film towards a higher RI value. In summary, we demonstrated the ARC of solar cells deposited by the PP method has advantages for higher PID demands.

4. Conclusions

The comparison of PID-resistance for the SiN_x films deposited by the direct and indirect PECVD was evaluated. After the PID test, the power conversion efficiency loss of the cell via indirect PECVD method was -33.82%, however, the loss of the cell via direct PECVD method was only -0.82%. Direct PECVD provides dense SiN_x films for increasing the PID-resistant ability. By taking the main advantage of the extremely high throughput of inline-type indirect PECVD, we demonstrated that the anti-PID ability can be much improved from -31.82% to -2.79% by introducing a thin oxide layer between indirect PECVD-deposited SiN_x film and Si through a simple pre-oxidation step. The oxide layer prevents sodium ions from accumulating at the silicon stacking faults and diminishes the possible shunting path.

Regarding to the uprising PID concern, we introduced a novel PP PECVD method for ARC deposition on PERC-type solar cells. The films made by the PP PECVD method were investigated by dilute HF etching and FTIR, and the corresponding cells were tested by PID treatments. The HF-etching rate of PP deposited SiN_x film is 2.5 nm/min, which is 2.6 times lower than that of CP deposited SiN_x film. A lower HF-etching rate indicates that the films made by the PP method contain less oxygen atoms and prevent chemical decomposition ascribed to oxidation of Na. The FTIR results consistently showed the lower oxygen content of the PP deposited SiN_x film and revealed that the PP deposited SiN_x films are thinner or have a smaller refractive index if compared to the CP deposited ones. As a result, we demonstrated the PID results of the cells with PP deposited SiN_x film were comparable to the ones with CP deposited SiN_x film, and the degradations of all cells satisfied the general criteria of the IEC 62804 specification. In addition, the involvement of pulsed plasma exhibits an extra degree of freedom to tune the film properties. Moreover, we expect the PP deposited SiN_x films, with further optimal film thickness and RI, would perform more aggressive PID-resistance than the conventional CP deposited ones.

Author Contributions: Conceptualization, C.-H.K.; Methodology and Data Curation, T.-W.K. and Y.-L.L.; Writing-Original Draft Preparation, T.-C.C.; Writing-Review & Editing, Z.-P.Y. and I.-S.Y.

Funding: This research was funded by Taiwan Ministry of Science and Technology, grant number MOST 106 2221-E-009-199 and MOST 107-2221-E-259-001-MY2.

Acknowledgments: The authors would like to thank the financial support from Ministry of Science and Technology, Taiwan. We also thank E-Ton Solar Tech. for the fabrication of Si solar cells and modules.

Conflicts of Interest: The authors declare no conflict of interest.

References

1. Águas, H.; Ram, S.K.; Araújo, A.; Gaspar, D.; Vicente, A.; Filonovich, S.A.; Fortunato, E.; Martins, R.; Ferreira, I. Silicon thin film solar cells on commercial tiles. *Energy Environ. Sci.* **2011**, *4*, 4620–4632. [CrossRef]
2. Águas, H.; Mateus, T.; Vicente, A.; Gaspar, D.; Mendes, M.J.; Schmidt, W.A.; Pereira, L.; Fortunato, E.; Martins, R. Thin film silicon photovoltaic cells on paper for flexible indoor applications. *Adv. Funct. Mater.* **2015**, *25*, 3592–3598. [CrossRef]
3. Masson, G.; Kaizuka, L. *Trends 2017 in Photovoltaic Applications*; IEA PVPS T1-32:2017; International Energy Agency (IEA): Paris, France, 2017.

4. Swanson, R.; Cudzinovic, M.; DeCeuster, D.; Desai, V.; Jürgens, J.; Kaminar, N.; Mulligan, W.; Rodrigues-Barbarosa, L.; Rose, D.; Smith, D.; et al. The surface polarization effect in high-efficiency silicon solar cells. In Proceedings of the 15th International Photovoltaic Science and Engineering Conference (PVSEC-15), Shanghai, China, 10–15 October 2005; pp. 410–413.

5. Pingel, S.; Frank, O.; Winkler, M.; Daryan, S.; Geipel, T.; Hoehne, H.; Berghold, J. Potential induced degradation of solar cells and panels. In Proceedings of the 35th IEEE Photovoltaic Specialists Conference, Honolulu, HI, USA, 20–25 June 2010; IEEE: Piscataway, NJ, USA, 2010; pp. 2817–2822.

6. Naumann, V.; Hagendorf, C.; Grosser, S.; Werner, M.; Bagdahn, J. Micro structural root cause analysis of potential induced degradation in c-Si solar cell. *Energy Procedia* **2012**, *27*, 1–6. [CrossRef]

7. Luo, W.; Khoo, Y.S.; Hacke, P.; Naumann, V.; Lausch, D.; Harvey, S.P.; Singh, J.P.; Chai, J.; Wang, Y.; Aberle, A.G.; et al. Potential-induced degradation in photovoltaic modules: A critical review. *Energy Environ. Sci.* **2017**, *10*, 43–68. [CrossRef]

8. Berghold, J.; Frank, O.; Hoehne, H.; Pingel, S.; Richardson, B.; Winkler, M. Potential induced degradation of solar cells and panels. In Proceedings of the 25th European Photovoltaic Solar Energy Conference and Exhibition, Valencia, Spain, 6–10 September 2010; pp. 3753–3759.

9. Koch, S.; Nieschalk, D.; Berghold, J.; Wendlandt, S.; Krauter, S.; Grunow, P. Potential induced degradation effects on crystalline silicon cells with various antireflective coatings. In Proceedings of the 27th European Photovoltaic Solar Energy Conference and Exhibition, Frankfurt, Germany, 24–28 September 2012; pp. 1985–1990.

10. Pop, S.C.; Schulze, R.; Wang, X.; Wang, H.; Nee, J.; Inns, D.; Meisel, A.; Kapur, J.; Antoniadis, H. Pathways towards advanced PID resistance of 1500 V. In Proceedings of the 31st European Photovoltaic Solar Energy Conference and Exhibition, Hamburg, Germany, 14–18 September 2015; pp. 15–19.

11. Mishina, K.; Ogishi, A.; Ueno, K. Investigation on PID-resistant anti-reflection coating for crystaline silicon solar cells. In Proceedings of the 28th European Photovoltaic Solar Energy Conference and Exhibition, Paris, France, 30 September–4 October 2013; pp. 1139–1143.

12. Nagel, H.; Saint-Cast, P.; Glatthaar, M.; Glunz, S. Inline processes for the stabilization of p-type crystalline Si solar cells against potential-induced degradation. In Proceedings of the 29th European Photovoltaic Solar Energy Conference and Exhibition, Amsterdam, The Netherlands, 22–26 September 2014; pp. 2351–2355.

13. Mehlich, H.; Decker, D.; Scheit, U.; Uhlig, M.; Frigge, S.; Runge, M.; Heinze, B.; Sperlich, H.P.; Mai, J.; Schlemm, H.; et al. A new method for high resistance against potential induced degradation. In Proceedings of the 27th European Photovoltaic Solar Energy Conference and Exhibition, Frankfurt, Germany, 24–28 September 2012; pp. 3411–3413.

14. Parsons, G.N.; Boland, J.J.; Tsang, J.C. Selective deposition and bond strain relaxation in silicon PECVD using time modulated silane flow. *Jpn. J. Appl. Phys.* **1992**, *31*, 1943–1947. [CrossRef]

15. Kudlacek, P.; Rumphorst, R.F.; van de Sanden, M.C.M. Accurate control of ion bombardment in remote plasmas using pulse-shaped biasing. *J. Appl. Phys.* **2009**, *106*, 073303. [CrossRef]

16. Liu, X.M.; Wang, Y.; Song, Y.H.; Wang, Y.N. Effect of pulse duration on characteristics of modulated radio-frequency $SiH_4/N_2/NH_3$ discharges. *Thin Solid Films* **2011**, *519*, 6951–6954. [CrossRef]

17. Kim, D.; Lee, S.; Kim, B.; Kang, B.J.; Kim, D. Duty ratio-controlled reflective property of silicon nitride films deposited at room temperature using a pulsed-PECVD at SiH_4–NH_3 plasma. *Curr. Appl. Phys.* **2011**, *11*, S43–S46. [CrossRef]

18. Watanabe, Y.; Shiratani, M.; Kubo, Y.; Ogawa, I.; Ogi, S. Effects of low-frequency modulation on RF discharge chemical vapor deposition. *Appl. Phys. Lett.* **1988**, *53*, 1263–1265. [CrossRef]

19. Viera, G.; Andújar, J.L.; Sharma, S.N.; Bertran, E. Nanopowder of silicon nitride produced in radio frequency modulated glow discharges from SiH_4 and NH_3. *Surf. Coat. Technol.* **1998**, *100*, 55–58. [CrossRef]

20. Kim, B.; Kim, S.; Seo, Y.H.; Kim, D.H.; Kim, S.J.; Jung, S.C. Surface morphology of SiN film deposited by a pulsed-plasma enhanced chemical vapor deposition at room temperature. *J. Nanosci. Nanotechnol.* **2008**, *8*, 5363–5366. [CrossRef] [PubMed]

21. *International Technology Roadmap for Photovoltaic (ITRPV)*, 9th ed.; VDMA: Frankfurt, Germany, 2018.

22. *TS IEC 62804-1 Photovoltaic (PV) Modules—Test Methods for the Detection of Potential-Induced Degradation—Part 1: Crystalline Silicon*; International Electrotechnical Commission (IEC): Geneva, Switzerland, 2015.

23. *Understanding Potential Induced Degradation*; ENG-PID-270-01 3/13; Advanced Energy Industries, Inc.: Fort Collins, CO, USA, 2013.

24. Hacke, P. Establishment of a PID pass/fail test for crystalline silicon modules by examining field performance for five years. In Proceedings of the 27th Workshop on Crystalline Silicon Solar Cells & Modules: Materials and Processes, Breckenridge, CO, USA, 30 July–2 August 2017.

25. Wan, Y.; McIntosh, K.R.; Thomson, A.F. Characterisation and optimisation of PECVD SiN$_x$ as an antireflection coating and passivation layer for silicon solar cells. *AIP Adv.* **2013**, *3*, 032113. [CrossRef]

26. van den Oever, P.J.; Kessels, W.M.M.; Hoex, B.; Bosch, R.C.M.; van Erven, A.J.M.; Pennings, R.L.J.R.; Stals, W.T.M.; Bijker, M.D.; van de Sanden, M.C.M. Plasma properties of a novel commercial plasma source for high-throughput processing of c-Si solar cells. In Proceedings of the 31st IEEE Photovoltaic Specialists Conference, Lake Buena Vista, FL, USA, 3–7 January 2005; IEEE: Piscataway, NJ, USA, 2005; pp. 1320–1323.

27. Naumann, V.; Lausch, D.; Hähnel, A.; Bauer, J.; Breitenstein, O.; Graff, A.; Werner, M.; Swatek, S.; Großer, S.; Bagdahn, J.; et al. Explanation of potential-induced degradation of the shunting type by Na decoration of stacking faults in Si solar cells. *Sol. Energy Mater. Sol. Cells* **2014**, *120*, 383–389. [CrossRef]

28. Lausch, D.; Naumann, V.; Graff, A.; Hahnel, A.; Breitenstein, O.; Hagendorf, C.; Bagdahn, J. Sodium out diffusion from stacking faults as root cause for the recovery process of potential-induced degradation. *Energy Procedia* **2014**, *55*, 486–493. [CrossRef]

29. Munro, R.G.; Dapkunas, S.J. Corrosion characteristics of silicon carbide and silicon nitride. *J. Res. Nat. Inst. Stand. Technol.* **1993**, *98*, 607–631. [CrossRef] [PubMed]

30. Mayer, M.Y.; Riley, F.L. Sodium-assisted oxidation of reaction-bonded silicon nitride. *J. Mater. Sci.* **1978**, *13*, 1319–1328. [CrossRef]

31. Lipiñski, M. Silicon nitride for photovoltaic application. *Arch. Mater. Sci. Eng.* **2010**, *46*, 69–87.

32. Blech, M.; Laades, A.; Ronning, C.; Schröter, B.; Borschel, C.; Rzesanke, D.; Lawerenz, A. Detailed study of PECVD silicon nitride and correlation of various characterization techniques. In Proceedings of the 24th European Photovoltaic Solar Energy Conference and Exhibition, Hamburg, Germany, 21–25 September 2009; pp. 507–511.

33. Kobayashi, S.I. IR spectroscopic study of silicon nitride films grown at a low substrate temperature using very high frequency plasma-enhanced chemical vapor deposition. *World J. Condens. Matter Phys.* **2016**, *6*, 287–293. [CrossRef]

34. Bugaev, K.O.; Zelenina, A.A.; Volodin, V.A. Vibrational spectroscopy of chemical species in silicon and silicon-rich nitride thin films. *Int. J. Spectrosc.* **2012**, *2012*, 281851. [CrossRef]

35. Dauwe, S. Low Temperature Surface Passivation of Crystalline Silicon and Its Applications on the Rear Side of Solar Cells. Ph.D. thesis, University of Hannover, Hannover, Germany, 2004.

36. Lelièvre, J.F.; Fourmond, E.; Kaminski, A.; Palais, O.; Ballutaud, D.; Lemiti, M. Study of the composition of hydrogenated silicon nitride SiN$_x$:H for efficient surface and bulk passivation of silicon. *Sol. Energy Mater. Sol. Cells* **2009**, *93*, 1281–1289. [CrossRef]

 *coatings*

Article

Facile Solution Spin-Coating SnO$_2$ Thin Film Covering Cracks of TiO$_2$ Hole Blocking Layer for Perovskite Solar Cells

Haiyan Ren, Xiaoping Zou *, Jin Cheng, Tao Ling, Xiao Bai and Dan Chen

Research Center for Sensor Technology, Beijing Key Laboratory for Sensor, Ministry of Education Key Laboratory for Modern Measurement and Control Technology, School of Applied Sciences, Jianxiangqiao Campus, Beijing Information Science & Technology University, Beijing 100101, China; yanh3100@gmail.com (H.R.); chengjin@bistu.edu.cn (J.C.); taoling8102@gmail.com (T.L.); aiaiyan521@gmail.com (X.B.); danchen630@gmail.com (D.C.)
* Correspondence: hsh@mail.bistu.edu.cn; Tel.: +86-10-6488-4673 (ext. 816)

Received: 4 August 2018; Accepted: 3 September 2018; Published: 6 September 2018

Abstract: The hole blocking layer plays an important role in suppressing recombination of holes and electrons between the perovskite layer and fluorine-doped tin oxide (FTO). Morphological defects, such as cracks, at the compact TiO$_2$ hole blocking layer due to rough FTO surface seriously affect performance of perovskite solar cells (PSCs). Herein, we employ a simple spin-coating SnO$_2$ thin film solution to cover cracks of TiO$_2$ hole blocking layer for PSCs. The experiment results indicate that the TiO$_2$/SnO$_2$ complementary composite hole blocking layer could eliminate the serious electrical current leakage existing inside the device, extremely reducing interface defects and hysteresis. Furthermore, a high efficiency of 13.52% was achieved for the device, which is the highest efficiency ever recorded in PSCs with spongy carbon film deposited on a separated FTO-substrate as composite counter electrode under one sun illumination.

Keywords: perovskite solar cell; hole blocking layer; solution spin-coating; TiO$_2$/SnO$_2$ layer

1. Introduction

Currently, perovskite solar cells (PSCs) have been recognized as the most promising alternatives to conventional silicon solar cells due to high conversion efficiency, low manufacturing cost, and simple process [1–7]. As for typical planar heterojunction PSCs composed of working electrode (cathode)/hole blocking layer/perovskite layer/electron blocking layer/counter electrode (anode), the hole blocking layer plays an important role in extracting electrons from the perovskite layer and, as well as, suppressing recombination of holes and electrons between the perovskite layer and FTO [8–11]. A uniform, pinhole-free and well electrically conductive hole blocking layer is highly demanded for well performed planar heterojunction PSCs [8,12]. Liao et al. found that the pinholes on the surface of the single SnO$_2$ hole blocking layer coating on the rough FTO substrate were still hard to eliminate. Therefore, radio-frequency magnetron sputtering TiO$_2$ fully covered the surface of the FTO electrode, thus passivating the FTO surface defects and reducing the recombination, finally utilizing TiO$_2$/SnO$_2$ bilayer as hole blocking layer to improve device performance [8,13]. Furthermore, TiO$_2$ compact layer has commonly been used as the hole blocking layer in the planar heterojunction PSCs. Since the spin-coating process formed a highly irregular thick TiO$_2$ layer on top of the rough FTO layer, the pinholes appeared on the surface of the TiO$_2$ hole blocking layer, so that the perovskite light-absorber layer directly contacted the FTO layer, resulting in recombination of holes and electrons between the perovskite light-absorber layer and FTO layer. Though uniform and pinhole-free compact TiO$_2$ layers were prepared using atomic layer deposition to improve device performance, the atomic layer deposition method had the disadvantage of including a low crystallinity of thin film due to the low deposition temperature [14]. However, Choi et al. mitigated the influence of the morphological defects (such as an irregular film thickness and poor physical contact between the TiO_2 and the FTO layers) at the TiO_2 interface due to a rough FTO by implementing a TiO_2 hole blocking layer based on the anodization method [15]. Though the research by Choi et al. mitigated the influence of the morphological defects, the preparation technology was very complicated. Up to now, several research groups have developed carbon counter electrodes to reduce the costs of perovskite devices [16–18]. In our research group, previously, the PSC based on sponge carbon/FTO composite counter electrode obtained the power conversion efficiency of 10.7% [19].

In this work, we employ TiO_2/SnO_2 complementary composite layer prepared by facile solution spin-coating SnO_2 on compact TiO_2-coated rough FTO as the hole blocking layer for planar heterojunction PSCs to solve morphological defects, such as cracks, at the compact TiO_2 hole blocking layer due to rough FTO surface. However, due to TiO_2 hole blocking layer has relative lower electron mobility, leading to unbalanced carrier transport in the device, thus regular planar structure PSC usually has a strong hysteresis behavior [20]. Meanwhile, SnO_2 is currently recognized as the promising hole blocking layer substitution for the conventional TiO_2 due to high electron mobility, low conduction band minimum, and low-temperature preparation [20–25]. Grätzel et al. and co-workers reported planar PSCs with efficiencies close to 21% using a simple, solution-processed technological approach for depositing SnO_2 layers [26]. Thereby, we obtained the modified hole blocking layer thin film without cracks through solution spin-coating SnO_2 on compact TiO_2-coated rough FTO at 3000 rpm. Correspondingly, this TiO_2/SnO_2 thin film and component optimization of perovskite precursor solution contributed to the formation of high-quality perovskite thin films (such as high crystallinity, large grain size, low defect density, and good flatness). Notably, the largest size of individual perovskite grain reached 2.67 μm. Therefore, PSC based on the TiO_2/SnO_2 complementary composite hole blocking layer achieved high power conversion efficiencies (PCE) of 13.52% in size of 0.2 cm^2 under absolute ambient condition, which is the highest efficiency ever recorded in PSCs with such spongy carbon/FTO composite counter electrode. These improvements indicate that the TiO_2/SnO_2 complementary composite hole blocking layer could eliminate the serious electrical current leakage existing inside the device and reduce interface defects.

2. Materials and Methods

2.1. Materials

Fluorine-doped SnO_2 (FTO) substrates were obtained from Yingkou Opv Tech New Energy Co., Ltd. (Yingkou, China, 7–8 Ω/square, 2.2 mm in thickness, 1.5 × 1.5 cm^2 in specification). *N,N*-dimethylformamide (DMF) and dimethyl sulfoxide (DMSO) were purchased from Sa'en Chemical Technology (Shanghai, China) Co., Ltd. Acidic titanium dioxide solution (bl-TiO_2) was purchased from Shanghai MaterWin New Materials Co., Ltd. (Shanghai, China, product code of MTW-CL-H-002, commodity name of HH-TiO_x, colorless and transparent in appearance, 99.98% in purity). The SnO_2 colloid precursor (Cas No. 18282-10-5) was obtained from Alfa Aesar (Shanghai, China, tin (IV) oxide, 15% in H_2O colloidal dispersion). Before use, the particles were diluted by deionized water to 3%. Lead(II) Iodide (PbI_2, Cas No. 10101-63-0, yellow crystalline powder in appearance, purity >99.99%), Methylammonium iodide (CH_3NH_3I, MAI, Cas No. 14965-49-2, white powder in appearance, purity ≥99.5%), Formamidinium Iodide ($HC(NH_2)_2I$, FAI, Cas No. 879643-71-7, white powder in appearance, purity ≥99.5%), Methylammonium Bromide (CH_3NH_3Br, MABr, Cas No. 6876-37-5, white powder in appearance, purity ≥99.5%), Methylammonium Chloride (CH_3NH_3Cl, MACl, Cas No. 593-51-1, white powder in appearance, purity ≥99.5%), and commercial 2,2′,7,7′-tetrakis-(*N,N*-dip-methoxyphenylamine)-9,9′-spirobifluorene solution (Spiro-OMeTAD, Cas No. 207739-72-8, yellow powder in appearance, purity ≥99.5%) were purchased from Xi'an Polymer Light Technology Corp. (Xi'an, China).

2.2. Device Fabrication

The surface treatment of the FTO for complete coverage before deposition of any film which affect severely the film deposition and properties. Malviya et al. demonstrated that rigorous cleaning process yielded clean FTO surface, and rigorous cleaning of the substrates prior to the hematite deposition was crucial for achieving highly reproducible results [27]. In order to ensure the quality of the subsequent film deposition, we thoroughly cleaned the FTO glass substrate before proceeding to the next step. FTO substrate was firstly wiped using a mixed solution of detergent and deionized water. The resultant FTO substrates were ultrasonically cleaned with glass water (deionized water: acetone: 2-propanol = 1:1:1) and ethanol for 20 min, separately. The FTO glasses were dried with dryer for 30 min and then cleaned with ultraviolet ozone for 10 min. Compact TiO_2 layer was deposited on the cleaned FTO substrate by spin-coating an acidic titanium dioxide solution at 2000 rpm for 60 s. The substrate was annealed on a hotplate at 100 °C for 10 min, and then sintered in muffle furnace at 500 °C for 30 min. For TiO_2/SnO_2 complementary composite hole blocking layer based devices, due to the hydrophobicity of the TiO_2 thin film layer and the complete aqueous solution of the SnO_2 precursor solution, the FTO/TiO_2 substrates were firstly cleaned by ultraviolet ozone for 10 min in order to form a hydrophilic group on the surface of the film. Then the SnO_2 functional layer was prepared by spin-coating 3 wt % SnO_2 precursor solution on the compact TiO_2-coated FTO substrate for 30 s at 2000, 3000, and 4000 rpm, respectively [23]. Subsequently, the $FTO/TiO_2/SnO_2$ based substrates were dried on a hotplate at 150 °C for 30 min, and then cleaned by ultraviolet ozone for 10 min again to improve its hydrophilicity.

The perovskite thin film was deposited by a two-step spin-coating method [19]. Firstly, the prepared $FTO/TiO_2/SnO_2$ substrates were preheated on a hotplate at 70 °C. The precursor solution of PbI_2 was prepared by dissolving 0.5993 g PbI_2 powder in 1 mL mixture of DMF/DMSO (0.95:0.05 of volume ratio), and then spin-coated on the preheated $FTO/TiO_2/SnO_2$ substrates at 1500 rpm for 30 s. After that, the doped FAI precursor solution was prepared by mixing 60 mg FAI and 6 mg MABr and 6 mg MACl in 1 mL isopropanol, and then spin-coated on the PbI_2 layer at 1300 rpm for 30 s. In order to obtain a high-quality perovskite light-absorber layer, the spin-coated substrates were annealed on a hotplate at 150 °C for 15 min under ambient condition. The electron blocking layer was deposited on top of the perovskite layer by 3000 rpm for 30 s using Spiro-OMeTAD solution.

Finally, some cleaned FTO glasses were used as substrates to collect soot of a burning candle as spongy carbon counter electrodes. The spongy carbon/FTO composite counter electrode was then pressed on the Spiro-OMeTAD layers of uncompleted devices. The whole process is carried out in air condition with a relative humidity of 10%–20% at room temperature. The entire preparation process was described in Scheme 1.

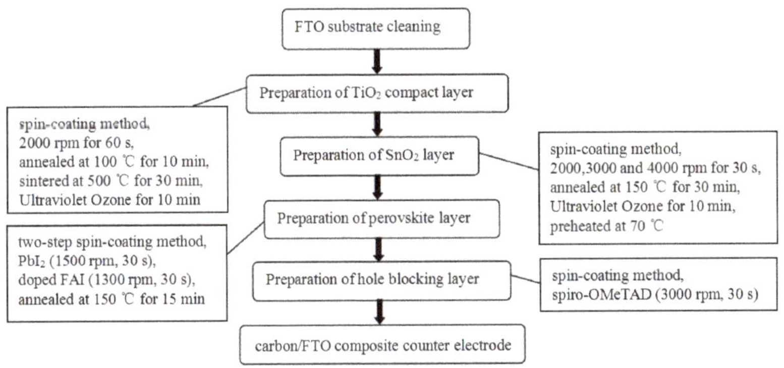

Scheme 1. The entire process of the device fabrication.

2.3. Characterization

X-ray diffraction (XRD) data from samples of perovskite films deposited on $FTO/TiO_2/SnO_2$ substrates were collected using an X-ray diffractometer (D8 Focus, Bruker, Dresden, Germany). The morphology of the perovskite films were measured using a scanning electron microscope (SEM) (SIGMA, Zeiss, Jena, Germany). The photo-current density–voltage (J–V) characteristics were measured under simulated standard air-mass AM 1.5 sunlight with using a solar simulator (Sol 3A, Oriel, Newport, RI, USA). All the measurements of the PSCs were performed under ambient atmosphere at room temperature without encapsulation.

3. Results and Discussion

Figure 1 shows the top-view SEM images of the bare FTO glass substrate (Figure 1a) and 3D roughness reconstruction of bare FTO glass (Figure 1f). As we can see, the surface of the bare FTO is very rough which corresponds to Sa (arithmetic-mean-height) of 74 nm. Many cracks (marked with a red square in Figure 1b) are observed on the surface of TiO_2 thin film. Figure 1c is the enlarged SEM surface image of the red square (in Figure 1b), however, the bottom FTO can be seen through the cracks, which will cause a large number of photocarriers to be recombination and seriously affect the performance of the PSC device. As shown in the cross-sectional SEM image (Figure 1d), the poor uniformity of TiO_2 thin film and the obvious fracture (marked with a red circle in Figure 1d) which was due to surface roughness of FTO, reveals that the bare FTO surface was exposed and could form direct contact with the perovskite layer, resulting in serious recombination of holes and electrons between the perovskite layer and FTO, in agreement with the results obtained with top-view SEM images (Figure 1b,c). The cross-sectional SEM image of bare FTO layer is shown in Figure 1e, we can find that the surface of bare FTO layer was very rough due to the presence of sharp peaks. Even some peaks and valleys reached a height of ~200 nm. However, as can be seen from Figure 1d, the spin-coating process formed a highly irregular thick TiO_2 layer due to overfilling the valleys on the rough FTO surface with TiO_2. The average thickness of single-layer TiO_2 compact layer was ~100 nm, which indicated some FTO peaks higher than 100 nm could not be covered, leading to the appearance of cracks in single-layer TiO_2 thin films. This is consistent with the observation from SEM images (Figure 1b,c) of the hole blocking layer thin films prepared with single-layer TiO_2 compact layer deposited on rough FTO substrates. The PSC devices based on single-layer TiO_2 hole blocking layer and double-layer TiO_2 hole blocking layer are named as device $R1$ and device $R2$, respectively. Table 1 shows the summary of J–V characteristics of device $R1$ and device $R2$ from reverse scan. As we can see, device $R1$ achieved low power conversion efficiency of 4.63% due to the presence of cracks. Compared with the device $R1$, the performance of the device $R2$ was improved. Huang et al. reported a facile dip-coating route to sequentially prepare a dense sol-gel-derived titanium dioxide (TiO_2) hole blocking layer and a smooth organolead halide perovskite film for pseudo-planar heterojunction perovskite solar cells [28]. Masood et al. used a scalable and low-cost dip coating method to prepare uniform and ultra-thin (5—50 nm) compact TiO_2 films on FTO glass substrates [29]. Though the dip-coating process does not require any complicated equipment and greatly minimizes the waste of source materials, enabling fast and easy production of uniform thin films, the method is not the same as our spin coating process.

Table 1. Summary of J–V characteristics of device $R1$ and device $R2$ from reverse scan.

Devices	N [a]	J_{sc} [b] (mA/cm^2)	V_{oc} [c] (V)	FF [d]	PCE [e] (%)
$R1$	1	13.13	0.86	41.09	4.63
$R2$	2	18.26	0.97	54.70	9.66

Notes: [a] Number of TiO_2 spin-coating layers; [b] Short-circuit photocurrent density; [c] Open-circuit voltage; [d] Fill factor; [e] Power conversion efficiency.

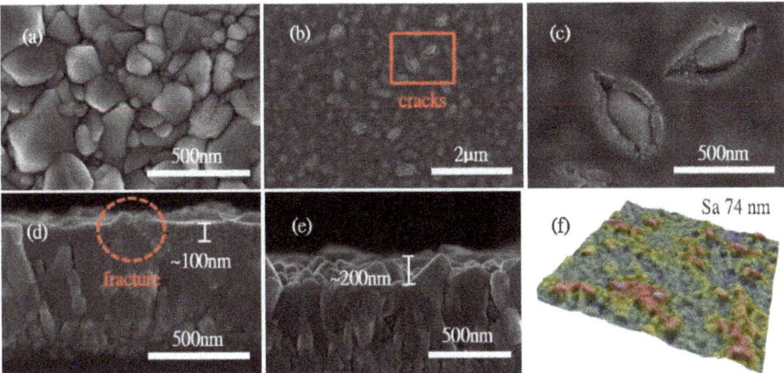

Figure 1. SEM images: (**a**) Top-view SEM image of bare FTO glass; (**b,c**) Top-view SEM image of TiO$_2$ film; (**d**) Cross-sectional SEM image of TiO$_2$ film; (**e**) Cross-sectional SEM image of bare FTO glass; (**f**) Top-view SEM image of 3D roughness reconstruction of bare FTO glass.

Figure 2 presents the top-view SEM images of SnO$_2$ thin films deposited on compact TiO$_2$-coated rough FTO for 30 s at 2000 rpm (Figure 2a,b), 3000 rpm (Figure 2c,d), and 4000 rpm (Figure 2e,f), respectively. As shown in Figure 2a,b, the cracks on the surface of the TiO$_2$/SnO$_2$ thin film were obviously filled with SnO$_2$ crystal grains. Due to the rotational speed being too low, a thick SnO$_2$ thin film formed. Thus, many cracks appeared on the surface of the SnO$_2$ thin film. Jiang et al. demonstrated a dense, pinhole-free film formed by spin coating SnO$_2$ nanoparticles solution onto glass/Indium-Tin Oxide (ITO) substrate at 3000 rpm [23]. However, in our experiment, as shown in Figure 2c,d, modified TiO$_2$/SnO$_2$ thin film without visible cracks through solution spin-coating SnO$_2$ on compact TiO$_2$-coated rough FTO at 3000 rpm formed, and is flat, uniform, and dense, which indicates the TiO$_2$/SnO$_2$ complementary composite hole blocking layer is able to fully cover the rough surface of the FTO electrode. Choi et al. demonstrated the TiO$_2$ film was so too thin that pinholes were observed in the TiO$_2$ layer, and electrons in the FTO layer easily recombined with holes due to the lack of hole blocking [15]. However, in our experiment, as the rotating speed increased to 4000 rpm, the bare FTO was exposed from a few large cracks in Figure 2e,f, which indicated that the thickness of the functional layer of SnO$_2$ was thinner due to the excessive spin-coating speed, and coverage of FTO/TiO$_2$ based substrate is incomplete. This result suggests that the direct contact between bare FTO and the perovskite light-absorber layer can cause the recombination of holes and electrons, leading to the degradation of device performance. It is consistent with the results obtained with summary of photovoltaic parameters with the J–V characteristics from reverse scan of the PSC devices based on TiO$_2$/SnO$_2$ complementary composite hole blocking layer.

However, the reasons for formation of cracks are different under high speed (4000 rpm) and low speed (2000 rpm), presented in detail in Figure 3. Figure 3 presents schematics for formation process of SnO$_2$ thin films at different rotational speed. Compared with situations of other rotational speed, the SnO$_2$ thin film from 2000 rpm was thicker, and the total amount of solution in the film was more. The solvent volatilization rate is lower during the spin coating process of 30 s, so that the solution film of SnO$_2$ was formed. During the annealing process of the solution film at 150 °C for 30 min, the solvent evaporated and the solution film contracted in the thickness direction, resulting in the exposure of the sharp peak of substrate, thus forming the cracks, as shown as Figure 3, corresponding to the SEM images (Figure 2a,b).

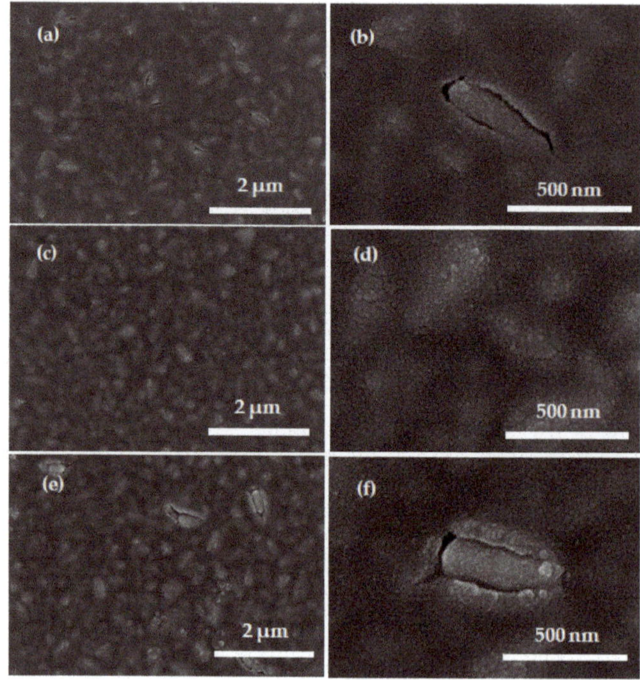

Figure 2. Top-view SEM images of SnO_2 thin films deposited on compact TiO_2-coated rough FTO at different spin-coating speeds for 30 s: (**a,b**) 2000 rpm; (**c,d**) 3000 rpm; (**e,f**) 4000 rpm.

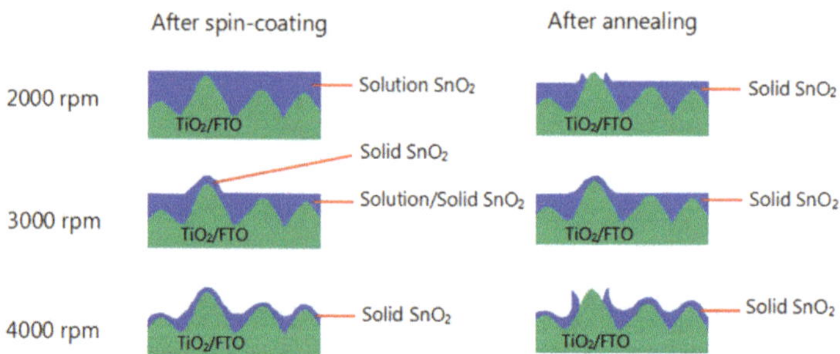

Figure 3. Schematics for formation of SnO_2 thin films at different rotational speed.

For the SnO_2 film from 3000 rpm, the solvent volatilized rate was more than the situation of SnO_2 film from 2000 rpm during the spin coating of 30 s, and the film itself was relatively thinner, thus forming a semi-solid/semi-liquid film after finishing the spin coating. In particular, the SnO_2 film at the location of the sharp peak of substrate almost became solid, thus preserving the shape at the location of the sharp peak, as shown as Figure 3, corresponding to the SEM images (Figure 2c,d).

In the process of spin-coating at 4000 rpm, the solvent volatilization was very fast, and the whole SnO_2 film was very thin and the total amount of solvent was very little, so the solid SnO_2 film had been formed after 30 s spin coating. It is worth noting that at the position of the sharp peak of substrate, the SnO_2 film was so thin that the ability to withstand strain of the film was very weak, resulting in a crack after annealing at 150 °C, as shown as Figure 3, corresponding to the SEM images (Figure 2e,f).

In addition, the film from spin coating at high speed was thinner than the one at low speed, and the film from spin coating at high speed was rougher than the one at low speed. Light trapping is a technique to improve optical absorption in the active layer of a solar cell. The rough surface can act as a light trapping structure to enhance the absorption perovskite solar cell [30–33]. However, the feature size here (<74 nm) is far less than the effective trapping sizes mentioned by previous research groups (~300 nm [30], ~6 µm [31] and ~800 nm [32]), so that the TiO_2/SnO_2 thin film with lower roughness is not the main reason for the efficiency drop.

Figure 4 presents the cross-sectional SEM images of the TiO_2/SnO_2 complementary composite hole blocking layer with SnO_2 spin-coated on compact TiO_2-coated FTO at 3000 rpm, which indicated the formation of modified TiO_2/SnO_2 film without visible breakages. As shown in Figure 4a, the entire film of TiO_2/SnO_2 complementary composite hole blocking layer was uniform in thickness and completely covered. Compared with single-layer TiO_2 hole blocking layer thin film, the SnO_2 thin film spin-coated on compact TiO_2-coated rough FTO exhibited a flatter surface. Notably, as shown in Figure 4b, the TiO_2/SnO_2 complementary composite hole blocking layer also has good coverage at the protrusions of the underlying substrate and no breakage or large bulging in the protrusions could be observed. This is consistent with the observation from top-view SEM images of SnO_2 films deposited on compact TiO_2-coated rough FTO at 3000 rpm (Figure 2c,d).

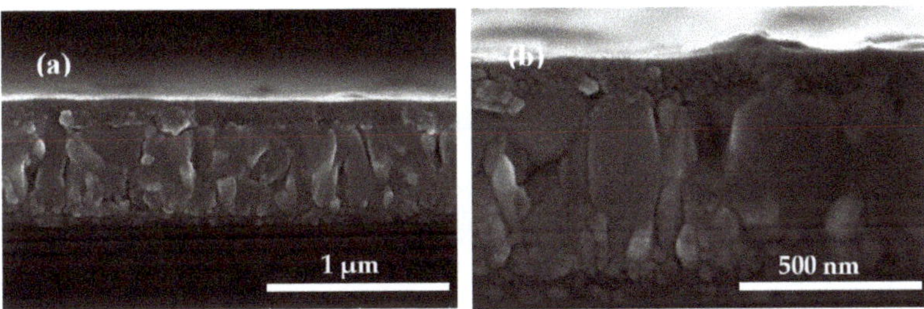

Figure 4. Cross-sectional SEM images of the TiO_2/SnO_2 complementary composite hole blocking layer with SnO_2 spin-coated on compact TiO_2-coated rough FTO at 3000 rpm: (**a**) Smaller magnification; (**b**) Larger magnification.

Figure 5 shows schematic illustration for the solution spin-coating process of TiO_2/SnO_2 complementary composite hole blocking layer. As we can see, the surface of the bare FTO was very rough due to the presence of sharp peaks (Figure 5a), corresponding to the SEM image (Figure 1e). As shown in Figure 5b, the spin-coating process formed a highly irregular thick TiO_2 layer due to overfilling the valleys on the rough FTO surface with TiO_2, resulting in many cracks appearing on the surface of TiO_2, and the bare FTO surface was exposed, corresponding to the SEM images (Figure 1b–d). However, as can be seen from Figure 5c, the cracks were absent from the surface of SnO_2 thin film deposited on compact TiO_2-coated rough FTO at 3000 rpm and the SnO_2 thin film fully covered the surface of the FTO electrode, corresponding to the SEM images (Figure 2c,d). Meanwhile, the SnO_2 thin film is flat, uniform, and dense, corresponding to the SEM images (Figure 4). This result suggests that the modified TiO_2/SnO_2 complementary composite hole blocking layer is beneficial to suppressing recombination of holes and electrons between the perovskite layer and FTO and enhances the performance of devices.

XRD data of the SnO_2 thin film deposited on FTO/TiO_2 substrate were collected in Figure 6. Each of the marked peak positions in the figure (Figure 6) corresponded to the (110), (101), (200), (211), (310), (301) crystal planes of the SnO_2 crystal, which was in consistent with the results of the previous report [23] about XRD pattern of the SnO_2 material, indicating SnO_2 thin film composite counter electrodes had been successfully deposited on the FTO/TiO_2 substrate.

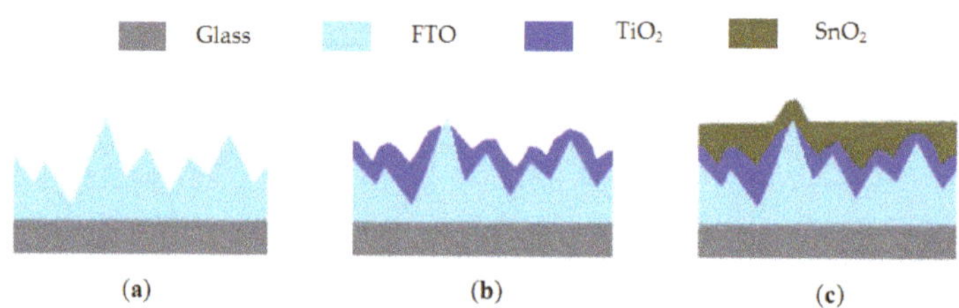

Figure 5. Schematic illustration for the solution spin-coating process of TiO$_2$/SnO$_2$ complementary composite hole blocking layer: (**a**) The bare FTO substrate; (**b**) TiO$_2$ thin film deposited on rough FTO substrate; (**c**) SnO$_2$ thin film deposited on compact TiO$_2$-coated rough FTO at 3000 rpm.

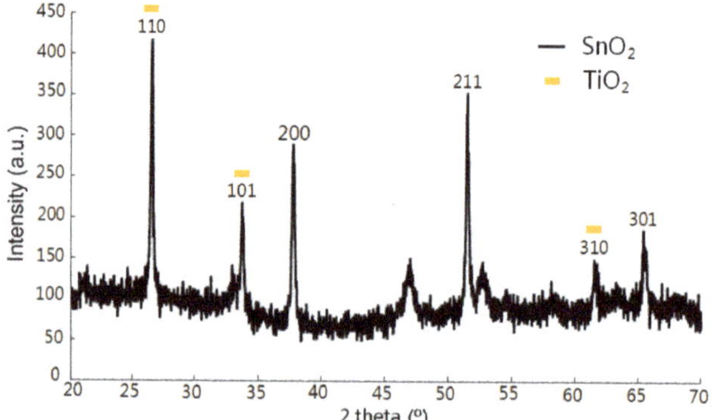

Figure 6. XRD pattern of SnO$_2$ thin film deposited on FTO/TiO$_2$ substrate.

Figure 7 presents top-view SEM images (Figure 7a,b) and cross-sectional SEM image (Figure 7c) of perovskite thin films deposited on TiO$_2$/SnO$_2$ of solution spin-coating SnO$_2$ on compact TiO$_2$-coated rough FTO at 3000 rpm, respectively. From Figure 7a,b, it can be seen that the entire perovskite film was high-quality and quite uniform free of cracks. However, high-quality perovskite film is essential for the performance of planar PSCs, as it can avoid the direct contact of hole blocking layer and electron blocking layer [20]. It is worth noting that the individual grain size of perovskite crystals was especially large, and the largest grain size reached a rare 2.67 μm. Furthermore, due to the passivation of the doping, the grain boundary was too smooth, which reduced defects of grain boundaries in the perovskite film to a great extent. From the cross-sectional SEM image (Figure 7c), it can be found that the entire perovskite film was composed of single layer of large grains and uniformly dense, which is coincident with the morphology of the film on the top-view SEM images (Figure 7a,b).

Figure 8 displays the XRD pattern of the perovskite thin film deposited on FTO/TiO$_2$/SnO$_2$ substrate in the case of spin-coating SnO$_2$ at 3000 rpm. Obviously, the XRD pattern presents strong and sharp (101) and (200) diffraction peaks at 14.02° and 28.32°, demonstrating the good crystallinity of perovskite film, which is consistent with the SEM results (Figure 7) of the large-grain-size perovskite film. The perovskite grains of good crystallinity effectively enhance the efficiency of injection of photo-carriers of the perovskite light-absorber layer into the hole blocking layer and electron blocking layer, thereby further improving the overall performance of the PSC device.

Figure 7. SEM images: (**a,b**) Top-view SEM images of FTO/TiO$_2$/SnO$_2$/perovskite; (**c**) Cross-sectional SEM image of FTO/TiO$_2$/SnO$_2$/perovskite.

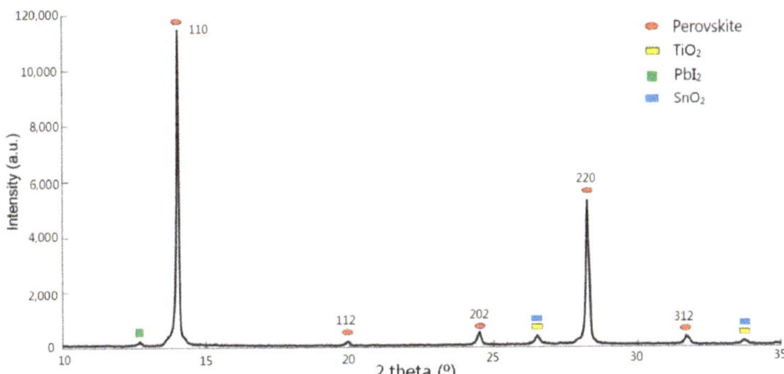

Figure 8. XRD pattern of the perovskite thin film deposited on FTO/TiO$_2$/SnO$_2$ substrate in the case of spin-coating SnO$_2$ at 3000 rpm.

Figure 9 shows energy bandgap diagram for TiO$_2$/SnO$_2$ based PSC, the shift of conduction band is not the main reason for the efficiency drop [34].

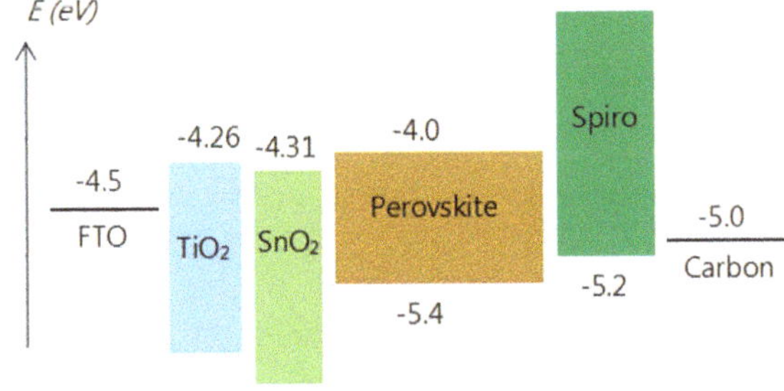

Figure 9. Energy bandgap diagram for TiO$_2$/SnO$_2$ based perovskite solar cells.

Figure 10 shows the *J–V* curves from reverse scan (RS) and forward scan (FS) for the PSC devices based on TiO_2/SnO_2 complementary composite hole blocking layer under different SnO_2 spin-coating speeds, measured under simulated sunlight with intensity of 100 mW cm^{-2} (AM 1.5 G). Obviously, when the SnO_2 spin-coating speed was 3000 rpm, the RS and FS curves of the device basically coincided and the hysteresis phenomenon basically disappeared, which matches the thin film morphology seen from the top-view SEM images (Figure 7a,b). When the SnO_2 spin-coating speeds were 2000 and 4000 rpm, a SnO_2 functional layer that was wither too thick or too thin could not match well with the TiO_2 compact layer to form modified TiO_2/SnO_2 complementary composite hole blocking layer. This led directly to a drop in the optoelectronic performance of the device. The presence of cracks also caused serious electrical current leakage existing inside the device, resulting the PSC devices under these two processes having different degrees of hysteresis.

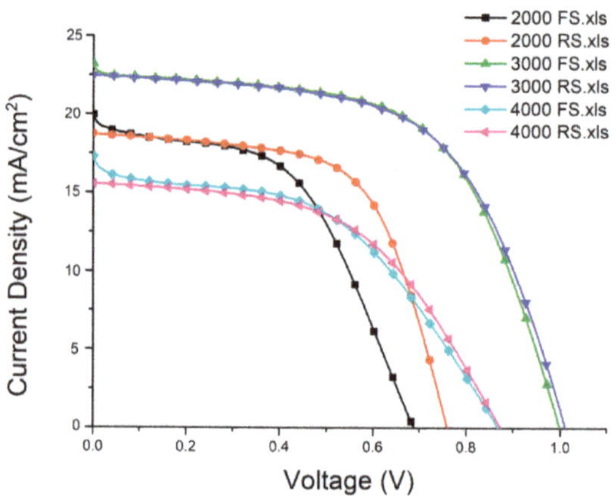

Figure 10. The *J–V* curves from reverse scan (RS) and forward scan (FS) for the PSC devices based on TiO_2/SnO_2 complementary composite hole blocking layer under different SnO_2 spin-coating speeds, measured under simulated sunlight with intensity of 100 mW cm^{-2} (AM 1.5 G).

Table 2 shows summary of *J–V* characteristics from RS of the PSC devices based on TiO_2/SnO_2 complementary composite hole blocking layer under different SnO_2 spin-coating speeds. The best cell based on TiO_2/SnO_2 complementary composite hole blocking layer under the SnO_2 spin-coating speed of 3000 rpm obtained a PEC of 13.52% corresponding to a J_{sc} of 22.48 mA cm^{-2}, a V_{oc} of 1.01 V and a FF of 59.51. Although this efficiency is not the highest in PSCs, it is the highest efficiency ever recorded in PSCs with such spongy carbon/FTO composite counter electrode, and furthermore, the preparation techniques of high-quality perovskite thin films and the cheap spongy carbon/FTO composite counter electrode under absolute ambient condition are beneficial for large-scale applications and commercialization. At the same time, since the problem of the defect at the interface between the hole blocking layer and the perovskite light-absorber layer was solved, the hysteresis phenomenon of the entire device was well solved.

Table 2. Summary of *J–V* characteristics from reverse scan of the PSC devices based on TiO_2/SnO_2 complementary composite hole blocking layer under different SnO_2 spin-coating speeds.

RSSC [a] (rpm)	J_{sc} [b] (mA/cm²)	V_{oc} [c] (V)	FF [d]	PCE [e] (%)
2000	18.74	0.76	61.95	8.81
3000	22.48	1.01	59.51	13.52
4000	15.56	0.87	52.22	7.10

Notes: [a] Rotating speed of spin-coating SnO_2; [b] Short-circuit photocurrent density; [c] Open-circuit voltage; [d] Fill factor; [e] Power conversion efficiency.

4. Conclusions

In summary, we optimized a rotational speed of spin-coating SnO_2 at 3000 rpm to cover the cracks of compact TiO_2-coated on rough FTO. Using composite layered TiO_2/SnO_2 can reduce hysteresis behavior due to an effectively suppressed recombination of holes and electrons between the perovskite layer and FTO. Meanwhile, as a result, we achieved high-quality perovskite thin films and high efficiency of 13.52% for devices with spongy carbon/FTO composite counter electrode. This work highlights the preparation process of modified hole blocking layer in PSC and has a certain reference significance for perovskite based opto-electronic devices.

Author Contributions: Conceptualization, X.B.; Methodology, X.Z.; Validation, D.C.; Investigation, H.R.; Data Curation, J.C.; Writing-Original Draft Preparation, H.R.; Writing-Review & Editing, T.L.; Visualization, H.R.; Supervision, X.Z.

Funding: This research was funded by Project of Natural Science Foundation of China (No. 61875186), Project of Natural Science Foundation of Beijing (No. Z160002), Opened Fund of the State Key Laboratory on Integrated Optoelectronics (No. IOSKL2016KF19), and Beijing Key Laboratory for Sensors of BISTU (No. KF20181077203).

Conflicts of Interest: The authors declare no conflicts of interest.

References

1. Snaith, H.J. Perovskites: The emergence of a new Era for low-cost, high-efficiency solar cells. *J. Phys. Chem. Lett.* **2013**, *4*, 3623–3630. [CrossRef]

2. Park, N.G. Organometal perovskite light absorbers toward a 20% efficiency low-cost solid-state mesoscopic solar cell. *J. Phys. Chem. Lett.* **2013**, *4*, 2423–2429. [CrossRef]

3. Green, M.A.; Ho-Baillie, A.; Snaith, H.J. The emergence of perovskite solar cells. *Nat. Photonics* **2014**, *8*, 506–514. [CrossRef]

4. Grätzel, M. The light and shade of perovskite solar cells. *Nat. Mater.* **2014**, *13*, 838–842. [CrossRef] [PubMed]

5. Abdi-Jalebi, M.; Pazoki, M.; Philippe, B.; Dar, M.I.; Alsari, M.; Sadhanala, A.; Diyitini, G.; Imani, R.; Lilliu, S.; Kullgren, J.; et al. Dedoping of lead halide perovskites incorporating monovalent cations. *ACS Nano* **2018**, *12*, 7301–7311. [CrossRef] [PubMed]

6. Alsari, M.; Bikondoa, O.; Bishop, J.; Abdi-Jalebi, M.; Ozer, L.Y.; Hampton, M.; Thompson, P.; Horantner, M.T.; Mahesh, S.; Greenland, C.; et al. In situ simultaneous photovoltaic and structural evolution of perovskite solar cells during film formation. *Energy Environ. Sci.* **2018**, *11*, 383–393. [CrossRef]

7. Alsari, M.; Pearson, A.J.; Wang, J.T.W.; Wang, Z.P.; Montisci, A.; Greenham, N.C.; Snaith, H.J.; Lilliu, S.; Friend, H. Degradation kinetics of inverted perovskite solar cells. *Sci. Rep.* **2018**, *8*, 5977. [CrossRef] [PubMed]

8. Liu, Z.Y.; Sun, B.; Liu, X.Y.; Han, J.H.; Ye, H.B.; Tu, Y.X.; Chen, C.; Shi, T.L.; Lang, Z.R.; Liao, G.L. 15% efficient carbon based planar-heterojunction perovskite solar cells using TiO_2/SnO_2 bilayer as electron transport layer. *J. Mater. Chem. A* **2018**, *6*, 7409–7419. [CrossRef]

9. Ren, Z.Q.; Wang, N.; Zhu, M.H.; Li, X.; Qi, J.Y. A NH_4F interface passivation strategy to produce air-processed high-performance planar perovskite solar cells. *Electrochim. Acta* **2018**, *282*, 653–661. [CrossRef]

10. Li, X.; Yang, J.Y.; Jiang, Q.H.; Lai, H.; Li, S.P.; Xin, J.W.; Chu, W.J.; Hou, J.D. Low-temperature solution-processed znse electron transport layer for efficient planar perovskite solar cells with negligible hysteresis and improved photostability. *ACS Nano* **2018**, *12*, 5605–5614. [CrossRef] [PubMed]

11. Ma, J.J.; Yang, G.; Qin, M.C.; Zheng, X.L.; Lei, H.W.; Chen, C.; Chen, Z.L.; Guo, Y.X.; Han, H.W.; Zhao, X.Z.; et al. MgO nanoparticle modified anode for highly efficient SnO_2-based planar perovskite solar cells. *Adv. Sci.* **2017**, *4*, 1700031. [CrossRef] [PubMed]

12. Wu, Y.Z.; Yang, X.D.; Chen, H.; Zhang, K.; Qin, C.J.; Liu, J.; Peng, W.Q.; Islam, A.; Bi, E.B.; Ye, F.; et al. Highly compact TiO_2 layer for efficient hole-blocking in perovskite solar cells. *Appl. Phys. Express* **2014**, *7*, 052301. [CrossRef]

13. Liu, Y.; Peng, Q.; Zhou, Z.P.; Yang, G. Effects of substrate temperature on properties of transparent conductive Ta-doped TiO_2 films deposited by radio-frequency magnetron sputtering. *Chin. Phys. Lett.* **2018**, *35*, 048101. [CrossRef]

14. Duong, T.T.; Kim, Y.J.; Eom, J.H.; Choi, J.S.; Le, A.T.; Yoon, S.G. Enhancement of solar cell efficiency using perovskite dyes deposited via a two-step process. *RSC Adv.* **2015**, *5*, 33515–33523. [CrossRef]

15. Choi, J.; Song, S.; Horantner, M.T.; Snaith, H.J.; Park, T. Well-defined nanostructured, single-crystalline TiO_2 electron transport layer for efficient planar perovskite solar cells. *ACS Nano* **2016**, *10*, 6029–6036. [CrossRef] [PubMed]

16. Zhang, N.; Guo, Y.; Yin, X.; He, M.; Zou, X. Spongy carbon film deposited on a separated substrate as counter electrode for perovskite-based solar cell. *Mater. Lett.* **2016**, *182*, 248–252. [CrossRef]

17. Mei, A.; Li, X.; Liu, L.; Ku, Z.; Liu, T.; Rong, Y.G.; Xu, M.; Hu, M.; Chen, J.; Yang, Y.; et al. A hole-conductor-free, fully printable mesoscopic perovskite solar cell with high stability. *Science* **2014**, *345*, 295–298. [CrossRef] [PubMed]

18. Habisreutinger, S.N.; Leijtens, T.; Eperon, G.E.; Stranks, S.D.; Nicholas, R.J.; Snaith, H.J. Carbon nanotube/polymer composites as a highly stable hole collection layer in perovskite solar cells. *Nano Lett.* **2014**, *14*, 5561–5568. [CrossRef] [PubMed]

19. Ling, T.; Zou, X.P.; Cheng, J.; Bai, X.; Ren, H.Y.; Chen, D. Modified sequential deposition route through localized-liquid-liquid-diffusion for improved perovskite multi-crystalline thin films with micrometer-scaled grains for solar cells. *Nanomaterials* **2018**, *8*, 416. [CrossRef] [PubMed]

20. Wu, C.C.; Huang, Z.R.; He, Y.H.; Luo, W.; Ting, H.K.; Li, T.Y.; Sun, W.H.; Zhang, Q.H.; Chen, Z.J.; Xiao, L.X. TiO_2/SnO_xCl_y double layer for highly efficient planar perovskite solar cells. *Org. Electron.* **2017**, *50*, 485–490. [CrossRef]

21. Ke, W.J.; Fang, G.J.; Liu, Q.; Xiong, L.B.; Qin, P.L.; Tao, H.; Wang, J.; Lei, W.H.; Li, B.R.; Wan, J.W.; et al. Low-temperature solution-processed tin oxide as an alternative electron transporting layer for efficient perovskite solar cells. *J. Am. Chem. Soc.* **2015**, *137*, 6730–6733. [CrossRef] [PubMed]

22. Song, J.X.; Zheng, E.Q.; Bian, J.; Wang, X.F.; Tian, W.J.; Sanehira, Y.; Miyasaka, T. Low-temperature SnO_2-based electron selective contact for efficient and stable perovskite solar cells. *J. Mater. Chem. A* **2015**, *3*, 10837–10844. [CrossRef]

23. Jiang, Q.; Zhang, L.Q.; Wang, H.L.; Yang, X.L.; Meng, J.H.; Liu, H.; Yin, Z.G.; Wu, J.L.; Zang, X.W.; You, J.B. Enhanced electron extraction using SnO_2 for high-efficiency planar-structure $HC(NH_2)_2PbI_3$-based perovskite solar cells. *Nat. Energy* **2017**, *2*, 16177. [CrossRef]

24. Baena, J.P.C.; Steier, L.; Tress, W.; Saliba, M.; Neutzner, S.; Matsui, T.; Giordano, F.; Jacobsson, T.J.; Kandada, A.R.S.; Zakeeruddin, S.M.; et al. Highly efficient planar perovskite solar cells through band alignment engineering. *Energy Environ. Sci.* **2015**, *8*, 2928–2934. [CrossRef]

25. Chen, J.Y.; Chueh, C.C.; Zhu, Z.L.; Chen, W.C.; Jen, A.K.Y. Low-temperature electrodeposited crystalline SnO_2 as an efficient electron transporting layer for conventional perovskite solar cells. *Sol. Energy Mater. Sol. Cells* **2017**, *164*, 47–55. [CrossRef]

26. Anaraki, E.H.; Kermanpur, A.; Steier, L.; Domanski, K.; Matsui, T.; Tress, W.; Saliba, M.; Abate, A.; Grätzel, M.; Hagfeldt, A.; et al. Highly efficient and stable planar perovskite solar cells by solution-processed tin oxide. *Energy Environ. Sci.* **2016**, *9*, 3128–3134. [CrossRef]

27. Malviya, K.D.; Dotan, H.; Yoon, K.R.; Kim, I.D.; Rothschild, A. Rigorous substrate cleaning process for reproducible thin film hematite (alpha-Fe_2O_3) photoanodes. *J. Mater. Res.* **2016**, *31*, 1565–1573. [CrossRef]

28. Huang, L.K.; Li, C.; Sun, X.X.; Xu, R.; Du, Y.Y.; Ni, J.; Cai, H.K.; Li, J.; Hu, Z.Y.; Zhang, J.J. Efficient and hysteresis-less pseudo-planar heterojunction perovskite solar cells fabricated by a facile and solution-saving one-step dip-coating method. *Org. Electron.* **2017**, *40*, 13–23. [CrossRef]

29. Masood, M.T.; Weinberger, C.; Sarfraz, J.; Rosqvist, E.; Sanden, S.; Sandberg, O.J.; Vivo, P.; Hashmi, G.; Lund, P.D.; Osterbacka, R.; et al. Impact of film thickness of ultrathin dip-coated compact TiO_2 layers on the performance of mesoscopic perovskite solar cells. *ACS Appl. Mater. Interfaces* **2017**, *9*, 17906–17913. [CrossRef] [PubMed]

30. Du, Q.G.; Shen, G.S.; John, S. Light-trapping in perovskite solar cells. *AIP Adv.* **2016**, *6*, 065002. [CrossRef]

31. Iftiquar, S.M.; Yi, J.S. Numerical simulation and light trapping in perovskite solar cell. *J. Photonics Energy* **2016**, *6*, 025507. [CrossRef]

32. Peer, A.; Biswas, R.; Park, J.M.; Shinar, R.; Shinar, J. Light management in perovskite solar cells and organic LEDs with microlens arrays. *Opt. Express* **2017**, *25*, 10704–10709. [CrossRef] [PubMed]

33.	Bhattacharya, J.; Peer, A.; Joshi, P.H.; Biswas, R.; Dalal, V.L. Blue photon management by inhouse grown ZnO:Al cathode for enhanced photostability in polymer solar cells. *Sol. Energy Mater. Sol. Cells* **2018**, *179*, 95–101. [CrossRef]

34.	Lee, Y.H.; Paek, S.; Cho, K.T.; Oveisi, E.; Gao, P.; Lee, S.; Park, J.S.; Zhang, Y.; Humphry-Baker, R.; Asiri, A.M.; et al. Enhanced charge collection with passivation of the tin oxide layer in planar perovskite solar cells. *J. Mater. Chem. A* **2017**, *5*, 12729–12734. [CrossRef]

Review

Review of CdTe$_{1-x}$Se$_x$ Thin Films in Solar Cell Applications

Martina Lingg *, Stephan Buecheler and Ayodhya N. Tiwari

Laboratory for Thin Films and Photovoltaics, Empa - Swiss Federal Laboratories for Materials Science and Technology, Ueberlandstrasse 129, 8600 Duebendorf, Switzerland
* Correspondence: martina.lingg@empa.ch

Received: 5 July 2019; Accepted: 13 August 2019; Published: 15 August 2019

Abstract: Recent improvements in CdTe thin film solar cells have been achieved by using CdTe$_{1-x}$Se$_x$ as a part of the absorber layer. This review summarizes the published literature concerning the material properties of CdTe$_{1-x}$Se$_x$ and its application in current thin film CdTe photovoltaics. One of the important properties of CdTe$_{1-x}$Se$_x$ is its band gap bowing, which facilitates a lowering of the CdTe band gap towards the optimum band gap for highest theoretical efficiency. In practice, a CdTe$_{1-x}$Se$_x$ gradient is introduced to the front of CdTe, which induces a band gap gradient and allows for the fabrication of solar cells with enhanced short-circuit current while maintaining a high open-circuit voltage. In some device structures, the addition of CdTe$_{1-x}$Se$_x$ also allows for a reduction in CdS thickness or its complete elimination, reducing parasitic absorption of low wavelength photons.

Keywords: CdTe; CdSe; CdTe$_{1-x}$Se$_x$; photovoltaics; solar cells; review

1. Introduction

With a global drive towards renewable energies, photovoltaics is an increasing part of global energy supply. CdTe is the leading technology amongst thin film solar cells for cost effective solar electricity production, due to its high photovoltaic conversion efficiency, long-term performance stability, low fabrication costs and short energy-payback time [1]. The current efficiency record for CdTe cells is 22.1%, held by First Solar [2,3].

In state-of-the-art CdTe solar cells, the J_{SC} is already close to its theoretical limit [4]. According to the Shockley–Queisser thermodynamic limit to solar cell efficiency, the optimum band gap for maximum theoretical solar cell efficiency under AM 1.5 G is 1.34 eV [5,6]; the band gap of CdTe is slightly too wide at 1.5 eV [4]. Strategies to shift the band gap of the CdTe absorber towards the theoretical optimum can help to further increase photocurrent and, consequently, the efficiency. CdTe$_{1-x}$Se$_x$ is a promising material for this purpose, as its band gap can be shifted down to 1.4 eV [7,8] by appropriately adjusting the composition. First Solar's world record cell efficiency was improved from 19.5% to 22.1% using an absorber with a lower band gap than CdTe, the CdTe$_{1-x}$Se$_x$ alloy [9–11].

With an absorber material that has an adjustable band gap, the band gap profile through the absorber layer can be engineered by applying a compositional grading. Different research groups have been investigating the use of a CdTe$_{1-x}$Se$_x$ layer towards the front of a CdTe absorber layer, resulting in champion devices with improved J_{SC} due to higher absorption in the low-band gap front layer, while maintaining the V_{OC} from the bulk CdTe [8,12–17].

For traditional CdS/CdTe solar cells, a CdCl$_2$ treatment, generally consisting of deposition of CdCl$_2$ on the absorber and subsequent heating, is needed to recrystallize the absorber layer, resulting in increased CdTe grain size and intermixing of CdS and CdTe as well as modifications in electronic properties leading to enhancement in photovoltaic conversion efficiency [18]. If a CdSe or CdTe$_{1-x}$Se$_x$ layer is inserted in between CdS and CdTe, interdiffusion during CdCl$_2$ treatment results in a CdTe$_{1-x}$Se$_x$

layer. Because $CdTe_{1-x}Se_x$ has no miscibility gap, a continuous composition gradient is formed, with an associated band gap gradient [15,19], allowing for absorption of high-wavelength photons. Several research groups have also found that, with the use of a $CdTe_{1-x}Se_x$ front layer, the CdS window layer thickness can be reduced, or the CdS layer can even be omitted completely, reducing parasitic absorption in the short-wavelength region and further increasing J_{SC} [8,14,20]. A thick CdS layer is generally required to avoid direct contact between the transparent conducting oxide (TCO) and the CdTe absorber, but a $CdTe_{1-x}Se_x$ absorber seems to be less sensitive to this issue. Further improvement was achieved by replacing CdS with MgZnO, which has a more favorable band alignment with both CdTe and $CdTe_{1-x}Se_x$ than CdS, and which is more transparent in the short-wavelength region [15,21,22]. Additionally, passivation of defects due to Se has been reported to be responsible for the higher carrier lifetimes in $CdTe_{1-x}Se_x$ compared to CdTe [19,23].

CdTe technology has shifted in recent years from a traditional CdS/CdTe structure to a device structure that utilizes a $CdTe_{1-x}Se_x$ layer and a band gap gradient, but no CdS layer, in order to achieve high currents and voltages. In this review paper we discuss the steps of this transformation, from the first introduction of a $CdTe_{1-x}Se_x$ layer to the complete replacement of the CdS layer by MgZnO and $CdTe_{1-x}Se_x$. We discuss the material properties of $CdTe_{1-x}Se_x$ and their dependence on composition and the application of this material in thin-film solar cells.

2. Growth of $CdTe_{1-x}Se_x$ Thin Films and Material Properties

In this section, we will discuss the deposition of $CdTe_{1-x}Se_x$ films and their structural, optical and electronic properties.

$CdTe_{1-x}Se_x$ is a solid solution without a miscibility gap; the alloy is stable over the complete composition range $0 \leq x \leq 1$. Thin films of desired composition can be grown with a variety of methods, such as high-vacuum evaporation of the elements [24] or the alloy $CdTe_{1-x}Se_x$ [25,26], co-evaporation of CdTe and CdSe [8,27], close space sublimation of the alloy $CdTe_{1-x}Se_x$ [23,28] or co-sublimation of the components CdTe and CdSe [29], molecular beam epitaxy [30,31], hot wall deposition [32], electron beam deposition [7,33], or electrodeposition [34,35]. The properties of grown layers depend on the deposition method and the conditions used.

In solar cells, the main methods used for $CdTe_{1-x}Se_x$ formation are high-vacuum evaporation [8] or radio frequency sputtering [14,16,20] of a CdSe layer next to a CdTe layer, with interdiffusion during $CdCl_2$ treatment, or close space sublimation of a $CdTe_{1-x}Se_x$ layer next to a CdTe layer [15], which also interdiffuse during $CdCl_2$ treatment. Close space sublimation is performed at higher temperatures than high-vacuum deposition or sputtering, and in CdTe it results in very large grains. The lower deposition temperature of high-vacuum evaporation results in smaller as-deposited grains. However, recrystallization during $CdCl_2$ treatment results in overall comparable grain sizes between different deposition methods [36]. The properties of $CdTe_{1-x}Se_x$ are therefore expected to depend on deposition parameters as well as $CdCl_2$ treatment conditions.

2.1. Crystal Structure

CdTe crystallizes in the cubic zinc blende structure, while CdSe crystallizes in the hexagonal wurzite structure. The two structures are closely related; each Cd^{2+} ion is surrounded by a tetrahedron of four Se^{2-} and/or Te^{2-} ions. In the solid solution $CdTe_{1-x}Se_x$, both the zinc blende and the wurzite structure have been reported [16,34,37–39]. For $0.2 \leq x \leq 0.5$, cubic zinc blende is the low-temperature structure, with a transition to hexagonal wurzite at temperatures above 800 °C. This transition temperature decreases with increasing x, and at $x = 0.6$ the hexagonal wurzite structure is formed at 600 °C [16,37]. With appropriate deposition parameters, mainly dependent on substrate temperature, the entire composition range of $CdTe_{1-x}Se_x$ can be fabricated in the zinc blende structure. In this structure, the lattice constant a decreases with increasing x, following Vegard's law, due to the smaller size of Se^{2-} compared to Te^{2-} [7,33,40,41]. The wurzite structure can occur at room temperature for $x = 0.3$ and higher, sometimes mixed with the zinc blende structure [34,38,39]. In solar cell applications, the zinc blende structure is desired, as it is photoactive, i.e., it can convert light into photovoltaic current, while the wurzite structure is not photoactive [16].

2.2. Optical Properties

CdTe$_{1-x}$Se$_x$ has a direct band gap with a pronounced bowing behavior. Figure 1 shows reported band gaps of CdTe$_{1-x}$Se$_x$ thin films measured with different methods, as well as one band gap bowing derived from first-principles DFT calculations, all showing a similar bowing behavior trend. The bowing parameter b can be extracted from experimental data with a second-degree polynomial fit. The band gap of any composition x can then be calculated using Equation (1) [42]:

$$E_G(x) = x \cdot E_G^{CdSe} + (1-x) \cdot E_G^{CdTe} - b \cdot x \cdot (1-x) \tag{1}$$

Band gap bowing parameters from bulk and thin-film CdTe$_{1-x}$Se$_x$, fabricated by different methods and using different measurement techniques, are listed in Table 1. The published bowing parameter values cover a range of 0.56–0.97, and they all agree on the composition of the band gap minimum at $0.3 \leq x \leq 0.4$, independent of fabrication method. The bowing only marginally depends on the crystal structure [39]. The band gap determination using photoluminescence and cathodoluminescence seem to yield bowing parameters slightly higher than transmittance/reflectance measurements, but they are still in good agreement [31].

Theoretical calculations using different methods are also shown in Table 1, and they seem to be in good agreement with experimental data [42–44]. The band gap minima in all experiments and calculations are around 1.4 eV, confirming that CdTe$_{1-x}$Se$_x$ is a good candidate for band-gap engineering in CdTe solar cells for the purpose of increasing J_{SC}.

In CdTe$_{1-x}$Se$_x$, the valence band edge is dominated by Se and Te p-orbitals, and the conduction band edge is dominated by Cd s- and p-states. Therefore, the anion substitution would be expected to primarily affect the valence band edge [43,45]. However, first-principles calculations show that the bowing of the band gap is caused by the bowing of both band edges due to strong intra-band coupling in both the valence and the conduction band [44].

The refractive index of CdTe$_{1-x}$Se$_x$ is almost constant over a large range of wavelengths, as well as with changes in composition and film thickness. Use of a CdTe$_{1-x}$Se$_x$ layer is therefore not expected to change the overall reflectance of a solar cell, and the introduction of a CdTe$_{1-x}$Se$_x$ gradient does not introduce a gradient in the refractive index [7,25,32].

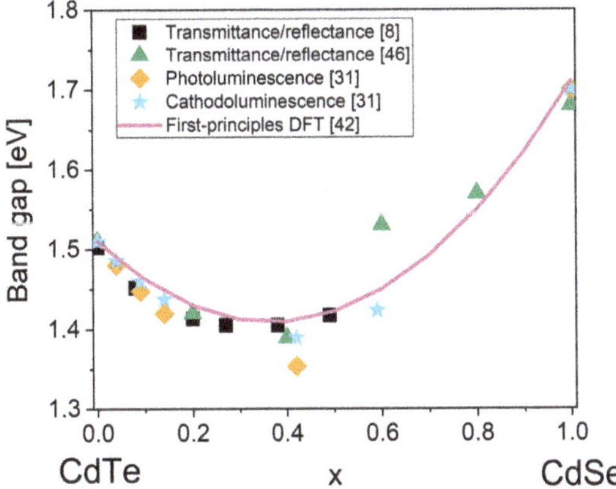

Figure 1. Band gap bowing of CdTe$_{1-x}$Se$_x$ measured by different methods [8,31,46] and calculated from first-principles DFT [42].

Table 1. Band gap bowing parameters of layers and powders synthesized with different methods.

Source	Deposition	Band Gap Determination	Crystal Structure	x Range	Bowing Parameter	x of Band Gap Minimum	Band Gap Minimum
[39]	Sintering of mixed CdTe and CdSe powders	Transmittance/reflectance	zinc blende	0–0.6	0.74	0.35	1.38
[39]	Sintering of mixed CdTe and CdSe powders	Transmittance/reflectance	wurzite	0.4–1	0.68	0.33	1.4
[8]	Co-evaporation of CdTe and CdSe	Transmittance/reflectance	zinc blende	0–0.5	0.78	0.37	1.39
[7]	Electron-beam evaporation	Transmittance/reflectance	zinc blende	0–1	0.567	0.37	1.44
[46]	Thermal evaporation of mixed CdTe and CdSe powders	Transmittance/reflectance	zinc blende	0–1	0.571	0.35	1.44
[31]	Molecular beam epitaxy on Si	Photoluminescence	zinc blende	0–1	0.97	0.40	1.35
[31]	Molecular beam epitaxy on Si	Cathodoluminescence	zinc blende	0–1	0.82	0.38	1.39
[42]	–	First-principles DFT calculations	zinc blende	0–1	0.75	0.40	1.41
[43]	–	sp^{3}s* tight-binding method calculations	zinc blende	0–1	0.904	0.39	1.37
[44]	–	First-principle hybrid-functional calculations	zinc blende	0–1	0.725	0.39	1.43

2.3. Electronic Properties

The electronic properties of $CdTe_{1-x}Se_x$ have not been studied extensively. As-deposited $CdTe_{1-x}Se_x$ is an n-type semiconductor with carrier densities between 10^{13} cm^{-3} (CdTe) and 10^{18} cm^{-3} (CdSe) [40]. The Hall mobility increases by less than an order of magnitude between CdTe and CdSe. Both carrier density and mobility are dependent on deposition temperature, with higher values at higher deposition temperatures due to increased grain size and possibly a reduction in structural defects [26,40]. In solar cell processing, the $CdTe_{1-x}Se_x$ layer is subjected to a $CdCl_2$ treatment, during which it recrystallizes. In CdTe, the recrystallization results in larger grains, reduced structural defects, and reduced recombination [18,47]. A similar effect is expected for $CdTe_{1-x}Se_x$. In absorbers consisting of a $CdTe_{1-x}Se_x$ layer in front of CdTe, recrystallization during $CdCl_2$ treatment has been shown to be responsible for a reduction in stacking faults, combined with interdiffusion of the $CdTe_{1-x}Se_x$ and CdTe layers [19,28].

In solar cells that use $CdTe_{1-x}Se_x$ as part of the absorber layer, dopants are generally added to the whole $CdTe_{1-x}Se_x$/CdTe stack and permeate the $CdTe_{1-x}Se_x$ layer via diffusion. A commonly used extrinsic dopant is copper, which forms deep acceptors as Cu on a Cd site (Cu_{Cd}) [48,49]. First-principles calculations indicate that the formation energy of this acceptor defect exhibits a bowing, with lower formation energies in the $CdTe_{1-x}Se_x$ solid solution than in either parent compound [44]. In the mixed alloy, the bonding Te orbitals are partially replaced with Se orbitals, which have a lower energy. The electronic effects of this change in bonding orbitals are one reason for the bowing behavior. Another reason is the smaller lattice constant with increasing x leading to reduced compressive strain when Cd is replaced by the much smaller Cu [44]. The reduced formation energy indicates better dopability of $CdTe_{1-x}Se_x$ with copper compared to CdTe. However, Hall effect measurements of $CdTe_{1-x}Se_x$ with $0 \leq x \leq 0.2$, deposited by high-vacuum evaporation and subjected to a $CdCl_2$ treatment, revealed that the achievable doping concentration is lower in Se-containing samples, with more than an order of magnitude difference between CdTe and $x = 0.2$ [8]. A possible explanation is that compensating donors, such as copper on interstitial sites (Cu_i), are formed as well, and that the formation energy of these defects is also dependent on x. If the electronic and/or structural effects of $CdTe_{1-x}Se_x$ alloying facilitate the formation of compensating donors more than the formation of acceptor defects, the result is a limit to the achievable hole density [8,49,50].

The photoluminescence decay of $CdTe_{1-x}Se_x$ measured by time-resolved photoluminescence (TRPL) is generally dominated by surface recombination, resulting in short apparent lifetimes [23,51]. In order to get a reliable measure for the bulk minority carrier lifetime, the surface has to be passivated, e.g., using Al_2O_3, which provides field-effect passivation [52]. Another approach is two-photon excitation (2PE) TRPL, which allows for lifetime measurements at a selected depth of the absorber [51,53]. In $CdTe_{1-x}Se_x$ layers ($x = 0.2$, deposited by close space sublimation and subjected to $CdCl_2$ treatment) with both surfaces passivated by sputtered Al_2O_3, lifetimes of up to 430 ns were measured, an order of magnitude higher than lifetimes in similarly passivated CdTe [23]. The reason seems to be that Se passivates a deep defect in the bulk of the absorber, even at low concentrations [19]. This is also correlated with longer diffusion length in material with higher x [13,19]. Two-photon excitation TRPL measurements performed on an interdiffused $CdTe_{1-x}Se_x$/CdTe absorber (close space sublimation, $CdCl_2$ treated, resulting maximum Se content $x \approx 0.16$) have shown lifetimes of around an order of magnitude higher in the Se-rich front layer compared to the CdTe back layer. This lifetime difference within a single absorber was linked to reduced recombination at the grain boundaries in the Se-containing region [51,53]. Se therefore, at least up to $x \approx 0.2$, seems to provide both passivation of a bulk defect and of the grain boundaries, resulting in an overall increased lifetime. Further studies are needed to accurately determine the lifetime dependence with x, and to determine the steepness of the lifetime gradient in an interdiffused $CdTe_{1-x}Se_x$/CdTe absorber.

The various deposition methods used for CdSe or $CdTe_{1-x}Se_x$ apply a variety of substrate temperatures and deposition rates. We expect that the reported properties of $CdTe_{1-x}Se_x$ may depend on deposition method as well as on the parameters of the $CdCl_2$ treatment, but further investigations are needed to determine the extent.

3. $CdTe_{1-x}Se_x$ in Solar Cells

In recent years, $CdTe_{1-x}Se_x$ has been used as a layer at the front of a predominantly CdTe absorber layer (see Figure 2), either deposited as CdSe or as $CdTe_{1-x}Se_x$, which interdiffuse with CdTe during $CdCl_2$ treatment to form a continuous composition gradient. This results in an absorber layer with a low band gap, increasing the J_{SC}, and it allows for a reduction in CdS thickness, both in superstrate and substrate configuration. As a further step, Munshi et al. [15] replaced the CdS with MgZnO for the highest device efficiencies. Schematics of the different device configurations in which $CdTe_{1-x}Se_x$ layers have been used are shown in Figure 2, from CdS/$CdTe_{1-x}Se_x$/CdTe (Figure 2b), to $CdTe_{1-x}Se_x$/CdTe (Figure 2c), to MgZnO/$CdTe_{1-x}Se_x$/CdTe (Figure 2d). As an alternative to the conventional superstrate configuration, a device in substrate configuration (Figure 2e) is discussed as well.

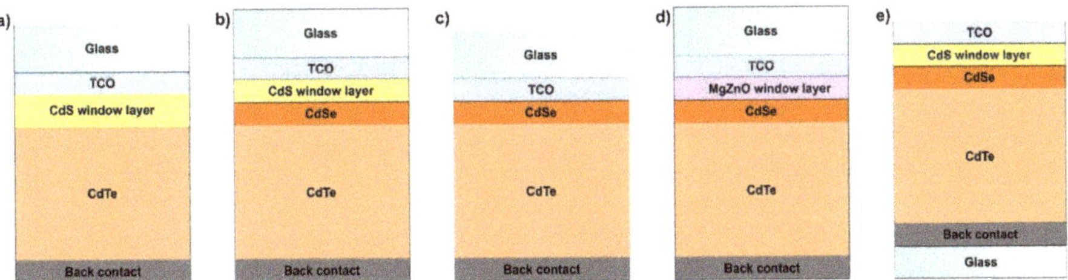

Figure 2. Schematics of deposited layers for solar cells (intermixing not depicted, thicknesses not to scale) for (**a**) a conventional CdTe device, (**b**) a device with a CdSe layer between CdTe and CdS [20]; (**c**) a device with CdSe replacing CdS [14,16]; (**d**) a device with MgZnO and CdSe replacing CdS [15]; and (**e**) a device in substrate configuration with a CdSe layer between CdTe and CdS [8].

The best reported devices for each of the different structures in Figure 2 are listed in Table 2, as well as the CdTe$_{1-x}$Se$_x$-containing record device by First Solar. By removing the CdS layer, a small increase in J_{SC} and FF has been shown, due to reduced parasitic absorption of CdS, at the cost of some losses in V_{OC}, due to interface recombination at the TCO/CdTe$_{1-x}$Se$_x$ interface [20]. With a MgZnO window layer, all device parameters could be improved: high J_{SC} due to the absence of parasitic absorption in the window layer, and high V_{OC} and FF due to a good band alignment between MgZnO and CdTe$_{1-x}$Se$_x$, resulting in a power conversion efficiency of 19.1% [15]. For the substrate configuration device, while the FF is comparable to superstrate devices, the V_{OC} and J_{SC} are lower. The loss in V_{OC} may be due to interface recombination at the CdS/CdTe$_{1-x}$Se$_x$ interface; the loss in J_{SC} due to parasitic absorption in the remaining CdS layer [8]. The world record device from First Solar surpasses all other devices in terms of V_{OC} and J_{SC}, and only the MgZnO/CdTe$_{1-x}$Se$_x$ device can exceed its FF. The exact structure and therefore the reasons for this outstanding performance are not public knowledge, but the high J_{SC} and V_{OC} indicate the use of MgZnO or a similar window layer [3,9,10].

Table 2. *J-V* parameters of the best CdTe$_{1-x}$Se$_x$-containing devices for each configuration shown in Figure 2.

Source	Device Structure	V_{OC} [mV]	J_{SC} [mA/cm^2]	FF [%]	Efficiency [%]
Paudel and Yan [20]	Figure 2b	806	27.2	64.1	14.1
Paudel and Yan [20]	Figure 2c	771	27.5	69.4	14.7
Munshi et al. [15]	Figure 2d	854	28.4	79.1	19.1
Lingg et al. [8]	Figure 2e	710	25.6	67	12.2
First Solar [3,9,10]	unknown	887.2	31.69	78.5	22.1

In the following section, we will first discuss the deposition methods used to form a CdTe$_{1-x}$Se$_x$/CdTe solar cell absorber, and in subsequent subsections we will discuss the details of how CdTe$_{1-x}$Se$_x$ has been used to improve performance in devices with different configurations.

3.1. Formation of a CdTe$_{1-x}$Se$_x$ Layer during Device Processing

Different deposition methods have been used to form a CdTe$_{1-x}$Se$_x$ layer. In conventional superstrate configuration, CdSe can be deposited by high-vacuum evaporation [27], magnetron sputtering [14,16,47], or pulsed laser deposition [54], with CdTe$_{1-x}$Se$_x$ formation during CdCl$_2$ treatment. An alternative is close-space sublimation (CSS) of a CdTe$_{1-x}$Se$_x$ layer [15,19,29]. The CdSe or CdTe$_{1-x}$Se$_x$ layer is followed by deposition of CdTe, CdCl$_2$ and doping treatments, and back contact deposition.

For devices in substrate configuration, a CdSe layer is deposited onto CdTe, followed by CdCl$_2$ and doping treatments and window layer/front contact deposition (see Figure 2e) [8].

In all of these device structures, the absorber and window layer are recrystallized and the CdSe/CdTe$_{1-x}$Se$_x$ interdiffuse with the CdTe during CdCl$_2$ treatment. In an as-deposited CdTe$_{1-x}$Se$_x$ and CdTe absorber, a large number of small grains towards the front (CdTe$_{1-x}$Se$_x$) and increasing grain size towards the back are observed, but after the CdCl$_2$ treatment, the grains are uniform in size and no distinct boundary is visible between CdTe$_{1-x}$Se$_x$ and CdTe [28]. Scanning transmission electron microscopy (STEM)/energy-dispersive x-ray spectroscopy (EDS) mapping reveals interdiffusion of CdTe$_{1-x}$Se$_x$ and CdTe, with Se diffusing more than a µm into the CdTe layer [15,28]. Secondary ion mass spectrometry (SIMS) reveals accumulation of Se at grain boundaries in the CdTe layer, and a relative loss of Se at grain boundaries in the CdTe$_{1-x}$Se$_x$ layer, which is a strong indication for a combination of fast diffusion of Se along the grain boundaries and slow diffusion into the grains [19].

Any photons with energies above the CdTe band gap ($\geq$1.5 eV) are absorbed in the CdTe$_{1-x}$Se$_x$ layer or in the CdTe absorber underneath. CdTe$_{1-x}$Se$_x$ has a lower band gap than CdTe (down to 1.4 eV), and it therefore absorbs photons that cannot be absorbed by CdTe. To maximize J_{SC} gain, the CdTe$_{1-x}$Se$_x$ layer has to be sufficiently thick to absorb the majority of these photons with energies below the CdTe band gap. If it is formed from CdSe, the composition of the $CdTe_{1-x}Se_x$ layer after interdiffusion depends on the thickness of the deposited CdSe layer as well as on the annealing conditions. Yan et al. found that for CdSe layers up to 100 nm thick, the interdiffusion is sufficient to result in a completely photoactive zinc blende absorber with compositions of $x \leq 0.65$. For thicker layers, a non-photoactive wurzite layer with $x > 0.65$ remains at the front of the absorber, which is undesirable because of parasitic absorption. With this method, it is difficult to achieve a thick photoactive layer for full absorption without formation of a non-photoactive layer [16,20,47]. Forming the $CdTe_{1-x}Se_x$ layer by depositing $CdTe_{1-x}Se_x$ with $x \leq 0.65$ instead of CdSe allows for a higher overall Se content in the absorber because the risk of forming a non-photoactive wurzite layer is minimized, and thick photoactive low-band gap layers can be formed [15].

The interdiffusion of $CdTe_{1-x}Se_x$ and CdTe creates a compositional gradient, which is expected to be correlated to the gradient in the material properties of $CdTe_{1-x}Se_x$, such as band gap, lattice constant, charge carrier concentration, and minority carrier lifetime. In particular, Fiducia et al. [19] reported gradients in both band gap and diffusion length (determined from cathodoluminescence measurements) that are directly correlated with the Se concentration gradient in a $CdTe_{1-x}Se_x/CdTe$ device.

3.2. Overview of Device Performances with $CdTe_{1-x}Se_x$

Different research groups have had varying success in improving device performance using CdSe. Published device parameters are listed in Table 3, where the parameters are compared with reference devices without CdSe. Generally, the J_{SC} could be improved by the reduced band gap and by reducing or removing the CdS layer. V_{OC} losses are partially due to the reduced band gap. Further V_{OC} losses together with fill factor losses have been mostly attributed to interface recombination, but they were not further investigated. The highest V_{OC} was achieved by use of a MgZnO window layer instead of CdS. The deposition methods are also listed as they may have an influence on the material properties and device performance. The different approaches and results are discussed in the following subsections.

Table 3. *J-V* parameters of devices containing CdSe compared with non-CdSe devices. Deposition methods are abbreviated as RFS (radio frequency sputtering), CSS (close space sublimation), and HVE (high vacuum evaporation).

Source	Device Structure	V_{OC} [mV]	J_{SC} [mA/cm^2]	FF [%]	Efficiency [%]
Poplawsky et al. [16]	CdSe (100 nm, RFS)/CdTe (CSS)	770	27	60.2	12.6
	CdS (130 nm, RFS)/CdTe (CSS)	810	23.8	75.4	14.5
	difference with CdSe	−40	+3.2	−15.2	−1.9
Mia et al. [14]	CdSe (100 nm, RFS)/CdTe (CSS)	690	26.9	64.8	12.1
	CdS (140 nm, RFS)/CdTe (CSS)	847	24.7	70	14.6
	difference with CdSe	−157	+2.2	−5.2	−2.5
Paudel and Yan [20]	CdSe (100 nm RFS)/CdTe (CSS)	771	27.5	69.4	14.7
	CdS (15 nm, RFS)/CdSe (100 nm, RFS)/CdTe (CSS)	806	27.2	64.1	14.1
	CdS (130 nm, RFS)/CdTe (CSS)	811	24.2	75.5	14.8
	difference only CdSe	−40	+3.3	−6.1	−0.1
	difference mixed CdS/CdSe	−5	+3	−11.4	−0.7
pLingg et al. [8]	CdS (30 nm)/CdSe (60 nm)/CdTe (all HVE, substrate config.)	710	25.6	67	12.2
	CdS (120 nm)/CdTe (all HVE, substrate config.)	830	18.5	69.4	10.5
	difference with CdSe	−120	+7.1	−2.4	+1.7
Munshi et al. [15]	MgZnO (RFS)/CdTe$_{0.8}$Se$_{0.2}$ (800 nm, CSS)/CdTe (CSS)	854	28.4	79.1	19.1
	MgZnO (RFS)/CdTe (CSS)	860	26.3	78.9	17.9
	difference with CdSe	−6	+2.1	+0.2	+1.2
First Solar [3,9,10]	$CdTe_{1-x}Se_x$/CdTe *	887.2	31.69	78.5	22.1
	CdS/CdTe *	875.9	30.25	79.4	21
	difference with CdSe	+11.3	+1.44	−0.9	+1.1

* The exact device structures and methods of deposition used in the First Solar record devices are not public knowledge.

3.3. CdSe/CdTe Devices without a Window Layer

The performance of devices where the CdS window layer is completely substituted by a CdSe layer (Figure 2c) depends strongly on the thickness of the CdSe layer. Up to 100 nm thick CdSe layers completely interdiffuse with the CdTe during $CdCl_2$ treatment, forming photoactive zinc blende $CdTe_{1-x}Se_x$. This results in increased J_{SC} compared to CdS/CdTe devices (see Figure 3) on the one hand because of increased absorption in the long-wavelength region due to a reduced band gap of the absorber, and on the other hand because the absorbing CdS layer is eliminated (see Figure 4). The overall efficiency however is still reduced because of losses in V_{OC} and FF.

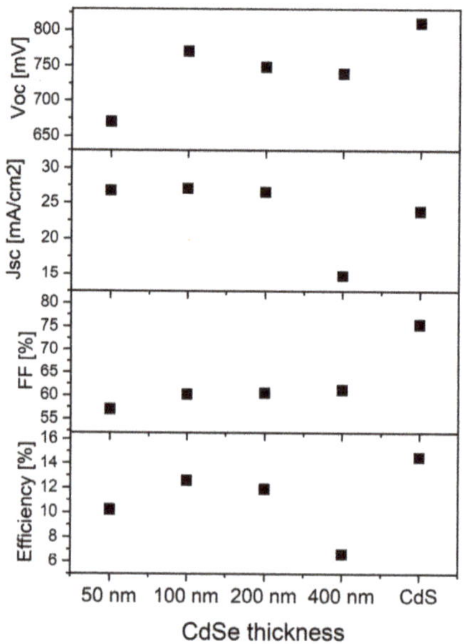

Figure 3. Performance of CdSe/CdTe solar cells with different CdSe thicknesses (values taken from Poplawsky et al. [16]).

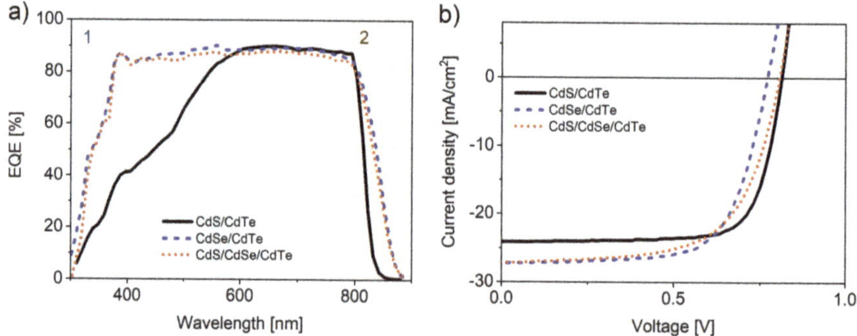

Figure 4. External quantum efficiency (EQE) (**a**) and *J-V* (**b**) curves of solar cells with different window layers (adapted from [20]).

With thicker CdSe layers, the interdiffusion is not complete, and the J_{SC} is reduced due to parasitic absorption in residual non-photoactive wurzite $CdTe_{1-x}Se_x$ with high Se content ($x > 0.65$). With optimized CdSe thickness, the J_{SC} can be increased by more than 3 mA/cm², however, this is not sufficient to offset the losses in V_{OC}, which are attributed to high interface recombination at the direct contact between $CdTe_{1-x}Se_x$ and TCO [16,20].

The TCO used in CdTe and $CdTe_{1-x}Se_x$/CdTe devices is usually a bilayer of a low-resistivity material, such as SnO_2:F (FTO), and a high-resistivity material, such as SnO_2. The highly resistive transparent (HRT) layer is used to prevent direct contact between the absorber and the FTO in the case of incomplete CdS coverage [55]. Baines et al. [12] investigated different oxides as HRT layers for $CdTe_{1-x}Se_x$ absorbers and found SnO_2 to be the best. It reduces V_{OC} losses that occur from direct contact of $CdTe_{1-x}Se_x$ with FTO, but the interface recombination is still higher than with a CdS layer [12,20]. In fact, the devices listed in Table 3 all use a SnO_2 HRT layer.

With electron-beam induced current (EBIC) measurements the position of the junction can be determined: for up to 100 nm CdSe, the junction is located at the $CdTe_{1-x}Se_x$/TCO interface. This reveals that the $CdTe_{1-x}Se_x$ does not act as a window layer, but as part of the p-side of the junction. For thicker CdSe layers, the junction is shifted into the $CdTe_{1-x}Se_x$ layer, to the interface between photoactive and non-photoactive $CdTe_{1-x}Se_x$ [16]. The V_{OC} and J_{SC} decrease for thicker layers, confirming that the presence of the non-photoactive wurzite $CdTe_{1-x}Se_x$ is not beneficial for the device performance. Deposition methods and parameters therefore have to be chosen to minimize the formation of non-photoactive $CdTe_{1-x}Se_x$.

3.4. CdSe/CdTe Devices with a CdS Window Layer

With a combined CdS/CdSe window layer (Figure 2b), device performance can be increased by combining the upsides of both layers while minimizing the drawbacks (see Table 3). By reducing the CdS thickness from 130 nm in a CdS/CdTe device to 15 nm in a CdS/CdSe/CdTe device, the parasitic absorption in the short-wavelength region is minimized (region 1 in Figure 4a) and the J_{SC} is increased. Further J_{SC} increase is due to a shift in band gap with $CdTe_{1-x}Se_x$ (region 2 in Figure 4a). The J_{SC} is therefore very similar to a CdSe/CdTe device, while the V_{OC} is improved with the CdS layer in between the $CdTe_{1-x}Se_x$ and the TCO, which reduces the interface recombination (see Figure 4b) [20,54].

A CdS/CdSe/CdTe structure also works in substrate configuration solar cells (Figure 2e). Substrate configuration CdTe solar cells require a thick CdS layer in order to mitigate the formation of pinholes. With a $CdTe_{1-x}Se_x$ layer between CdTe and CdS, the CdS layer thickness can be reduced from 120 to 30 nm with only small losses in V_{OC} and with a 7.1 mA/cm^2 increase in J_{SC} due to the reduction in CdS thickness and the reduction in band gap, resulting in an overall improvement in efficiency (see Table 3) [8].

3.5. $CdTe_{1-x}Se_x$/CdTe Devices with a MgZnO Window Layer

Recent high-efficiency CdTe devices have been achieved by replacing the CdS window layer with MgZnO (see Table 3) [15,22]. MgZnO has a band gap of 3.7 eV, higher than CdS, thereby reducing parasitic absorption, and it has a better band alignment with CdTe than TCO/CdS, reducing interface recombination and allowing for a higher V_{OC} [21,22]. However, introduction of a MgZnO window layer into a device structure can be problematic. The MgZnO/CdTe interface is sensitive to barriers originating from oxides, e.g., MgO, that can be formed during deposition. To obtain high-efficiency MgZnO/CdTe devices, deposition parameters, such as temperature and oxygen partial pressure during deposition, and $CdCl_2$ treatment, have to be carefully optimized [56].

Further improvement was achieved by using a $CdTe_{1-x}Se_x$ layer at the front of the CdTe absorber. With an 800 nm thick $CdTe_{1-x}Se_x$ layer in front of a CdTe absorber, Munshi et al. showed an improvement in device efficiency from 17.9% for a MgZnO/CdTe device to 19.1% in a MgZnO/$CdTe_{1-x}Se_x$/CdTe device (see Table 3) [15]. They improved the J_{SC} by 2.1 mA/cm^2 while retaining the high V_{OC} (see Figure 5a). As the device contains no CdS, the EQE in the short-wavelength region is improved (region 1 in Figure 5a).

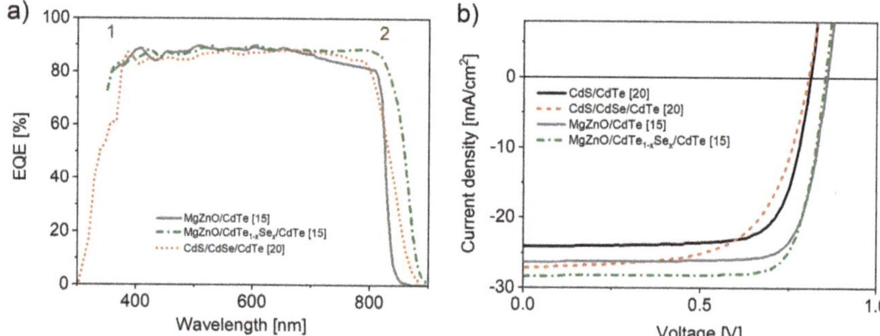

Figure 5. EQE (**a**) and *J-V* (**b**) curves of MgZnO/CdTe and MgZnO/CdTe$_{1-x}$Se$_x$/CdTe devices (adapted from [15]) compared with a CdS/CdSe/CdTe device (adapted from [20]).

The benefit of depositing CdTe$_{1-x}$Se$_x$ instead of CdSe in front of CdTe is the low possibility that non-photoactive wurzite CdTe$_{1-x}$Se$_x$ is formed. As a consequence, higher amounts of Se can be introduced to the device for better fine-tuning of the band gap gradient [29]. This results in a thicker photoactive low-band gap CdTe$_{1-x}$Se$_x$ layer in the finished device. The low-band gap region appears sufficiently thick to absorb most photons with energies below the band gap of CdTe, down to about 1.42 eV, as evidenced by the shift in absorption edge in the EQE curve in region 2 in Figure 5a. The difference in absorption between the thick photoactive layer achieved with CdTe$_{1-x}$Se$_x$ deposition and the thin photoactive layer achieved with CdSe deposition can be seen in the steepness of the EQE curve in region 2 in Figure 5a. The conclusion can be drawn that for maximized J_{SC} gain from CdTe$_{1-x}$Se$_x$, deposition of a CdTe$_{1-x}$Se$_x$ layer instead of a CdSe layer is necessary.

The high V_{OC} in the MgZnO/CdTe$_{1-x}$Se$_x$/CdTe device was in part attributed to a high lifetime of 22 ns measured by TRPL, four times higher than in the MgZnO/CdTe device, due to improved interface quality between MgZnO and CdTe$_{1-x}$Se$_x$, and also due to the passivation of a recombination center by Se [13,15,19]. The MgZnO/CdTe$_{1-x}$Se$_x$/CdTe device has a much higher V_{OC} than the CdS/CdTe$_{1-x}$Se$_x$/CdTe devices. This indicates that the V_{OC} in these structures is limited by the TCO/CdS/CdTe$_{1-x}$Se$_x$ interfaces, but further investigations into the V_{OC} and FF loss of CdTe$_{1-x}$Se$_x$/CdTe devices is needed to confirm this.

Figure 6 compares the EQE curve of the high-efficiency MgZnO/CdTe$_{1-x}$Se$_x$/CdTe device with the record device from First Solar. The First Solar device uses a CdTe$_{1-x}$Se$_x$ layer and some kind of buffer layer instead of CdS [11]. It has a higher V_{OC} and higher J_{SC} than the MgZnO/CdTe$_{1-x}$Se$_x$/CdTe device (see Table 3) [3,11]. The increase in J_{SC} is primarily due to a very good absorption at high wavelengths (region 2 in Figure 6). The very straight absorption edge indicates the presence of a sufficiently thick CdTe$_{1-x}$Se$_x$ layer with a band gap of 1.4 eV for complete absorption of photons in this region. Some further improvement of J_{SC} is achieved by an overall higher EQE, probably due to an antireflection coating. In the short-wavelength region (region 1 in Figure 6), the curves are very similar. This is an indication that First Solar must have used MgZnO or a similar material as a buffer layer, eliminating the influence of parasitic absorption. The increase in V_{OC} might be due to better interface engineering and therefore reduced recombination, but without more information on the First Solar device structure no conclusions can be drawn.

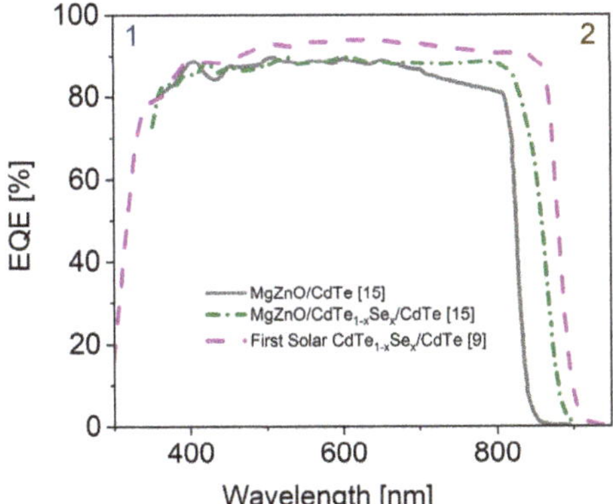

Figure 6. EQE curves of Munshi et al. (adapted from [15]) and the First Solar record device (adapted from [9]).

4. Conclusions

CdTe$_{1-x}$Se$_x$ has been successfully introduced into CdTe devices by several different groups. It has been shown that a band gap gradient can be achieved by several different techniques and deposition methods. For better control and a higher limit to the amount of Se put into the devices, use of a CdTe$_{1-x}$Se$_x$ layer is preferable to a CdSe layer in order to avoid formation of non-photoactive wurzite CdTeSe. In devices with a CdTe$_{1-x}$Se$_x$ layer at the front of the CdTe absorber, the reduced band gap and possibility of reduced CdS window layer thickness result in increased short-circuit current. The open circuit voltage can be maintained at the level of CdTe devices due to Se passivating a recombination center in the absorber. The best reported devices use a MgZnO window layer instead of CdS, which provides a better band alignment with the absorber and eliminates parasitic absorption from the window layer.

Funding: This research was funded by the Swiss National Science Foundation under the project 200021_160025/1.

Conflicts of Interest: The authors declare no conflict of interest.

References

1. Philipps, S.; Warmuth, W. *Photovoltaics Report*; Fraunhofer ISE: Freiburg, Germany, 2019.
2. First Solar Press Release. *First Solar Achieves Yet another Cell Conversion Efficiency World Record*; First Solar Press Release: Tempe, AZ, USA, 2016.
3. Green, M.A.; Emery, K.; Hishikawa, Y.; Warta, W.; Dunlop, E.D. Solar cell efficiency tables (version 48). *Prog. Photovolt. Res. Appl.* **2016**, *24*, 905–913. [CrossRef]
4. Geisthardt, R.M.; Topič, M.; Sites, J.R. Status and potential of CdTe solar-cell efficiency. *IEEE J. Photovolt.* **2015**, *5*, 1217–1221. [CrossRef]
5. Shockley, W.; Queisser, H.J. Detailed balance limit of efficiency of p-n junction solar cells. *J. Appl. Phys.* **1961**, *32*, 510. [CrossRef]
6. Rühle, S. Tabulated values of the Shockley–Queisser limit for single junction solar cells. *Sol. Energy* **2016**, *130*, 139–147. [CrossRef]
7. Islam, R.; Banerjee, H.; Rao, D. Structural and optical properties of CdSe$_x$Te$_{1-x}$ thin films grown by electron beam evaporation. *Thin Solid Films* **1995**, *266*, 215–218. [CrossRef]
8. Lingg, M.; Spescha, A.; Haass, S.G.; Carron, R.; Buecheler, S.; Tiwari, A.N. Structural and electronic properties of CdTe$_{1-x}$Se$_x$ films and their application in solar cells. *Sci. Technol. Adv. Mater.* **2018**, *19*, 683–692. [CrossRef]
9. Green, M.A.; Emery, K.; Hishikawa, Y.; Warta, W.; Dunlop, E.D. Solar cell efficiency tables (version 46). *Prog. Photovolt. Res. Appl.* **2015**, *23*, 805–812. [CrossRef]

10. Green, M.A.; Emery, K.; Hishikawa, Y.; Warta, W.; Dunlop, E.D. Solar cell efficiency tables (Version 45). *Prog. Photovolt.* **2015**, *23*, 1–9. [CrossRef]

11. Gang, X.; Gloeckler, M. High Efficiency CdTe Solar Cells and Modules. In Proceedings of the E-MRS Spring Meeting, Strasbourg, France, 18–22 June 2018.

12. Baines, T.; Zoppi, G.; Bowen, L.; Shalvey, T.P.; Mariotti, S.; DuRose, K.; Major, J.D. Incorporation of CdSe layers into CdTe thin film solar cells. *Sol. Energy Mater. Sol. Cells* **2018**, *180*, 196–204. [CrossRef]

13. Fiducia, T.A.M.; Munshi, A.H.; Barth, K.; Proprentner, D.; West, G.; Sampath, W.S.; Walls, J.M. Defect tolerance in As-deposited selenium-alloyed cadmium telluride solar cells. In Proceedings of the 2018 IEEE 7th World Conference on Photovoltaic Energy Conversion (WCPEC) (A Joint Conference of 45th IEEE PVSC, 28th PVSEC & 34th EU PVSEC), Waikoloa Village, HI, USA, 10–15 June 2018; pp. 0127–0130.

14. Mia, M.D.; Swartz, C.H.; Paul, S.; Sohal, S.; Grice, C.R.; Yan, Y.; Holtz, M.; Li, J.V. Electrical and optical characterization of CdTe solar cells with CdS and CdSe buffers—A comparative study. *J. Vac. Sci. Technol. B* **2018**, *36*, 052904. [CrossRef]

15. Munshi, A.H.; Kephart, J.; Abbas, A.; Raguse, J.; Beaudry, J.-N.; Barth, K.; Sites, J.; Walls, J.; Sampath, W. Polycrystalline CdSeTe/CdTe absorber cells with 28 mA/cm^2 short-circuit current. *IEEE J. Photovolt.* **2018**, *8*, 310–314. [CrossRef]

16. Poplawsky, J.D.; Guo, W.; Paudel, N.; Ng, A.; More, K.; Leonard, D.; Yan, Y. Structural and compositional dependence of the CdTe$_x$Se$_{1-x}$ alloy layer photoactivity in CdTe-based solar cells. *Nat. Commun.* **2016**, *7*, 12537.

17. Yang, X.; Liu, B.; Li, B.; Zhang, J.; Li, W.; Wu, L.; Feng, L. Preparation and characterization of pulsed laser deposited a novel CdS/CdSe composite window layer for CdTe thin film solar cell. *Appl. Surf. Sci.* **2016**, *367*, 480–484. [CrossRef]

18. McCandless, B.E.; Moulton, L.V.; Birkmire, R.W. Recrystallization and sulfur diffusion in CdCl$_2$-treated CdTe/CdS thin films. *Prog. Photovolt. Res. Appl.* **1997**, *5*, 249–260. [CrossRef]

19. Fiducia, T.A.M.; Mendis, B.G.; Li, K.; Grovenor, C.R.M.; Munshi, A.H.; Barth, K.; Sampath, W.S.; Wright, L.D.; Abbas, A.; Bowers, J.W.; et al. Understanding the role of selenium in defect passivation for highly efficient selenium-alloyed cadmium telluride solar cells. *Nat. Energy* **2019**, *4*, 504–511. [CrossRef]

20. Paudel, N.R.; Yan, Y. Enhancing the photo-currents of CdTe thin-film solar cells in both short and long wavelength regions. *Appl. Phys. Lett.* **2014**, *105*, 183510. [CrossRef]

21. Kephart, J.; McCamy, J.; Ma, Z.; Ganjoo, A.; Alamgir, F.; Sampath, W. Band alignment of front contact layers for high-efficiency CdTe solar cells. *Sol. Energy Mater. Sol. Cells* **2016**, *157*, 266–275. [CrossRef]

22. Munshi, A.H.; Kephart, J.M.; Abbas, A.; Shimpi, T.M.; Barth, K.L.; Walls, J.M.; Sampath, W.S. Polycrystalline CdTe photovoltaics with efficiency over 18% through improved absorber passivation and current collection. *Sol. Energy Mater. Sol. Cells* **2018**, *176*, 9–18. [CrossRef]

23. Kephart, J.M.; Kindvall, A.; Williams, D.; Kuciauskas, D.; Dippo, P.; Munshi, A.; Sampath, W.S. Sputter-deposited oxides for interface passivation of CdTe photovoltaics. *IEEE J. Photovolt.* **2018**, *8*, 587–593. [CrossRef]

24. Russak, M.A.; Creter, C. Vacuum Evaporated CdSe$_{1-x}$Te$_x$ Thin Films for Electrochemical Photovoltaic Cells. *J. Electrochem. Soc.* **1984**, *131*, 556–562. [CrossRef]

25. El-Nahass, M.; Sallam, M.; Afifi, M.; Zedan, I. Structural and optical properties of polycrystalline CdSe$_x$Te$_{1-x}$ ($0 \leq x \leq 0.4$) thin films. *Mater. Res. Bull.* **2007**, *42*, 371–384. [CrossRef]

26. Uthanna, S.; Reddy, P. Structural and electrical properties of CdSe$_x$Te$_{1-x}$ thin films. *Solid State Commun.* **1983**, *45*, 979–980. [CrossRef]

27. Borah, M.N.; Chaliha, S.; Sarmah, P.; Rahman, A. Electrical and optical properties of thin film (n) CdSe/(p) CdTe heterojunction and its performance as a photovoltaic converter. *J. Optoelectron. Adv. M.* **2008**, *10*, 1333–1339.

28. Munshi, A.H.; Kephart, J.M.; Abbas, A.; Danielson, A.; Gélinas, G.; Beaudry, J.-N.; Barth, K.L.; Walls, J.M.; Sampath, W.S. Effect of CdCl$_2$ passivation treatment on microstructure and performance of CdSeTe/CdTe thin-film photovoltaic devices. *Sol. Energy Mater. Sol. Cells* **2018**, *186*, 259–265. [CrossRef]

29. Swanson, D.E.; Sites, J.R.; Sampath, W.S. Co-sublimation of CdSe$_x$Te$_{1-x}$ layers for CdTe solar cells. *Sol. Energy Mater. Sol. Cells* **2017**, *159*, 389–394. [CrossRef]

30. Amir, F.; Clark, K.; Maldonado, E.; Kirk, W.; Jiang, J.; Ager III, J.; Yu, K.; Walukiewicz, W. Epitaxial growth of $CdSe_xTe_{1-x}$ thin films on Si (1 0 0) by molecular beam epitaxy using lattice mismatch graded structures. *J. Cryst. Growth* **2008**, *310*, 1081–1087. [CrossRef]

31. Campo, E.M.; Hierl, T.; Hwang, J.C.; Chen, Y.; Brill, G.; Dhar, N.K. Comparison of cathodoluminescence and photoluminescence of CdSeTe films grown on Si by molecular beam epitaxy. In Proceedings of the SPIE, Denver, CO, USA, 22 October 2004.

32. Muthukumarasamy, N.; Balasundaraprabhu, R.; Jayakumar, S.; Kannan, M.D.; Ramanathaswamy, P. Compositional dependence of optical properties of hot wall deposited $CdSe_xTe_{1-x}$ thin films. *Phys. Status Solidi (a)* **2004**, *201*, 2312–2318. [CrossRef]

33. Mangalhara, J.; Agnihotri, O.; Thangaraj, R. Structural, optical and photoluminescence properties of electron beam evaporated $CdSe_{1-x}Te_x$ films. *Sol. Energy Mater.* **1989**, *19*, 157–165. [CrossRef]

34. Bouroushian, M.; Loizos, Z.; Spyrellis, N.; Maurin, G. Influence of heat treatment on structure and properties of electrodeposited CdSe of Cd(Te, Se) semiconducting coatings. *Thin Solid Films* **1993**, *229*, 101–106. [CrossRef]

35. Kathalingam, A.; Kim, M.R.; Chae, Y.S.; Rhee, J.K.; Thanikaikarasan, S.; Mahalingam, T. Study on electrodeposited $CdSe_xTe_{1-x}$ semiconducting thin films. *J. Alloy. Compd.* **2010**, *505*, 758–761. [CrossRef]

36. McCandless, B.E.; Sites, J.R. *Handbook of Photovoltaic Science and Engineering*; John Wiley & Sons: Hoboken, NJ, USA, 2005; pp. 617–662.

37. Strauss, A.J.; Steininger, J. Phase diagram of the CdTe-CdSe pseudobinary system. *J. Electrochem. Soc.* **1970**, *117*, 1420–1426. [CrossRef]

38. Kumar, L.; Singh, B.P.; Misra, A.; Misra, S.; Sharma, T. Characterization of $CdSe_xTe_{1-x}$ sintered films for photovoltaic applications. *Phys. B Condens. Matter* **2005**, *363*, 102–109. [CrossRef]

39. Tai, H.; Nakashima, S.; Hori, S. Optical properties of $(CdTe)_{1-x}(CdSe)_x$ and $(CdTe)_{1-x}(CdS)_x$ systems. *Phys. Status Solidi (a)* **1975**, *30*, K115–K119. [CrossRef]

40. Russak, M.A. Deposition and characterization of $CdSe_{1-x}Te_x$ thin films. *J. Vac. Sci. Technol. A* **1985**, *3*, 433–435. [CrossRef]

41. Sanitarov, V.; Aleksandrova, L.; Kalinkin, I. The ranges of isomorphous substitutions in thin films of $CdSe_xTe_{1-x}$ solid solutions. *Thin Solid Films* **1982**, *97*, 205–214. [CrossRef]

42. Wei, S.H.; Zhang, S.; Zunger, A. First-principles calculation of band offsets, optical bowings, and defects in CdS, CdSe, CdTe, and their alloys. *J. Appl. Phys.* **2000**, *87*, 1304–1311. [CrossRef]

43. Tit, N.; Obaidat, I.M.; Alawadhi, H. Origins of bandgap bowing in compound-semiconductor common-cation ternary alloys. *J. Phys. Condens. Matter* **2009**, *21*, 075802. [CrossRef]

44. Yang, J.; Wei, S.-H. First-principles study of the band gap tuning and doping control in $CdSe_xTe_{1-x}$ alloy for high efficiency solar cell. *Chinese Phys. B* **2019**, *28*, 086106. [CrossRef]

45. Reshak, A.H.; Kityk, I.; Khenata, R.; Auluck, S. Effect of increasing tellurium content on the electronic and optical properties of cadmium selenide telluride alloys $CdSe_{1-x}Te_x$: An ab initio study. *J. Alloy. Compd.* **2011**, *509*, 6737–6750. [CrossRef]

46. Santhosh, T.; Bangera, K.V.; Shivakumar, G. Synthesis and band gap tuning in $CdSe_{1-x}Te_x$ thin films for solar cell applications. *Sol. Energy* **2017**, *153*, 343–347. [CrossRef]

47. Paudel, N.R.; Moore, K.L.; Yan, Y.; Poplawsky, J.D. Current enhancement of CdTe-based solar cells. *IEEE J. Photovolt.* **2015**, *5*, 1492–1496. [CrossRef]

48. Krasikov, D.; Knizhnik, A.; Potapkin, B.; Selezneva, S.; Sommerer, T. First-principles-based analysis of the influence of Cu on CdTe electronic properties. *Thin Solid Films* **2013**, *535*, 322–325. [CrossRef]

49. Wei, S.H.; Zhang, S.; Zhang, S.B. Chemical trends of defect formation and doping limit in II-VI semiconductors: The case of CdTe. *Phys. Rev. B* **2002**, *66*, 155211. [CrossRef]

50. Gretener, C.; Wyss, M.; Perrenoud, J.; Kranz, L.; Buecheler, S.; Tiwari, A.N. CdTe thin films doped by Cu and Ag-a comparison in substrate configuration solar cells. In Proceedings of the 2014 IEEE 40th Photovoltaic Specialist Conference (PVSC), Denver, CO, USA, 8–13 June 2014; pp. 3510–3514.

51. Kuciauskas, D.; Kanevce, A.; Burst, J.M.; Duenow, J.N.; Dhere, R.; Albin, D.S.; Levi, D.H.; Ahrenkiel, R.K. Minority carrier lifetime analysis in the bulk of thin-film absorbers using subbandgap (two-photon) excitation. *IEEE J. Photovolt.* **2013**, *3*, 1319–1324. [CrossRef]

52. Kuciauskas, D.; Kephart, J.M.; Moseley, J.; Metzger, W.K.; Sampath, W.S.; Dippo, P. Recombination velocity less than 100 cm/s at polycrystalline $Al_2O_3/CdSeTe$ interfaces. *Appl. Phys. Lett.* **2018**, *112*, 263901. [CrossRef]

53. Zheng, X.; Kuciauskas, D.; Moseley, J.; Colegrove, E.; Albin, D.S.; Moutinho, H.; Duenow, J.N.; Ablekim, T.; Harvey, S.P.; Ferguson, A.; et al. Recombination and bandgap engineering in CdSeTe/CdTe solar cells. *APL Mater.* **2019**, *7*, 071112. [CrossRef]
54. Yang, X.; Bao, Z.; Luo, R.; Liu, B.; Tang, P.; Li, B.; Zhang, J.; Li, W.; Wu, L.; Feng, L. Preparation and characterization of pulsed laser deposited CdS/CdSe bi-layer films for CdTe solar cell application. *Mater. Sci. Semicond. Process.* **2016**, *48*, 27–32. [CrossRef]
55. Feldman, S.; Mansfield, L.; Ohno, T.; Kaydanov, V.; Beach, J.; Nagle, T. Non-uniformity mitigation in CdTe solar cells: The effects of high-resistance transparent conducting oxide buffer layers. In Proceedings of the Conference Record of the Thirty-first IEEE Photovoltaic Specialists Conference, Lake Buena Vista, FL, USA, 3–7 January 2005; pp. 271–274.
56. Ablekim, T.; Perkins, C.; Zheng, X.; Reich, C.; Swanson, D.; Colegrove, E.; Duenow, J.N.; Albin, D.; Nanayakkara, S.; Reese, M.O.; et al. Tailoring MgZnO/CdSeTe Interfaces for Photovoltaics. *IEEE J. Photovolt.* **2019**, *9*, 888–892. [CrossRef]

Review

Encapsulation of Organic and Perovskite Solar Cells: A Review

Ashraf Uddin *, Mushfika Baishakhi Upama, Haimang Yi and Leiping Duan

School of Photovoltaic and Renewable Energy Engineering, University of New South Wales, Sydney 2052, Australia; m.upama@unsw.edu.au (M.B.U.); haimang.yi@student.unsw.edu.au (H.Y.); leiping.duan@unsw.edu.au (L.D.)
* Correspondence: a.uddin@unsw.edu.au

Received: 29 November 2018; Accepted: 21 January 2019; Published: 23 January 2019

Abstract: Photovoltaic is one of the promising renewable sources of power to meet the future challenge of energy need. Organic and perovskite thin film solar cells are an emerging cost-effective photovoltaic technology because of low-cost manufacturing processing and their light weight. The main barrier of commercial use of organic and perovskite solar cells is the poor stability of devices. Encapsulation of these photovoltaic devices is one of the best ways to address this stability issue and enhance the device lifetime by employing materials and structures that possess high barrier performance for oxygen and moisture. The aim of this review paper is to find different encapsulation materials and techniques for perovskite and organic solar cells according to the present understanding of reliability issues. It discusses the available encapsulate materials and their utility in limiting chemicals, such as water vapour and oxygen penetration. It also covers the mechanisms of mechanical degradation within the individual layers and solar cell as a whole, and possible obstacles to their application in both organic and perovskite solar cells. The contemporary understanding of these degradation mechanisms, their interplay, and their initiating factors (both internal and external) are also discussed.

Keywords: organic solar cells; perovskite solar cells; encapsulation; stability

---

1. Introduction

Organic photovoltaic (OPV) and Perovskite solar cell (PSC) are promising emerging photovoltaics thin film technology. Light harvester metal-halide perovskite materials, such as methyl-ammonium lead iodide ($CH_3NH_3PbI_3$), have exhibited small exciton binding energy, high optical cross-section, superior ambipolar charge transport, tuneable band gaps, and low-cost fabrication [1]. The $CH_3NH_3PbI_3$ PSCs and OPV devices can be solution-processed, which is suitable for roll-to-roll manufacturing processes for inexpensive largescale commercialization. The power conversion efficiency (PCE) of OPV devices have overpassed 14% for single junction and 17% for tandem devices to date with the development of low band-gap organic materials and device processing technology [2–5]. The achievement of the highest PCE of PSCs over 23.3% has shown promising future directions for using in large-scale production, together with traditional silicon solar cells [6]. Low-temperature (<150 °C) and solution-processed ZnO based electron transport layer (ETL) is one of the most promising materials for large scale roll-to-roll fabrication of perovskite and organic solar cells, owing to its almost identical electron affinity (4.2 eV) of TiO_2. Arafat et al. have already demonstrated aluminium (Al) doped ZnO (AZO) ETL of PSC with cell efficiency over 18% [7]. Currently, the poor stability of OPV and perovskite solar cells is a barrier for the commercialisation [8]. It is believed that oxygen and moisture are the external main reason for degradation of organic and PSCs, as shown in Figure 1 [9]. All the internal possible degradation mechanisms in PSCs can be controlled by careful interface engineering, such as a good choice of cathode and anode interlayer materials, ion-hybridizations in perovskite layer, etc.

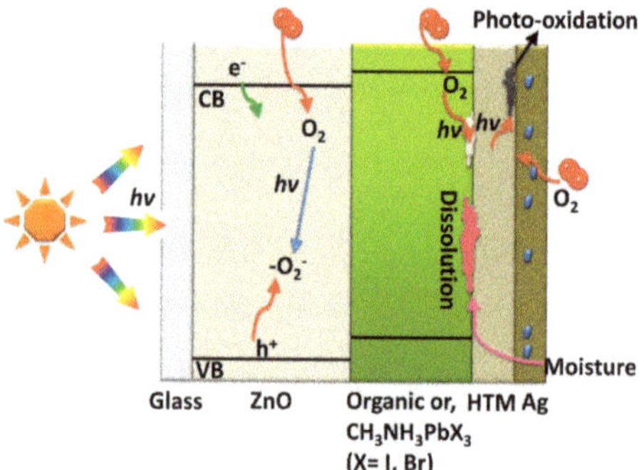

Figure 1. Schematic diagram of an organic/Perovskite solar cell (PSC) solar cell structure. The electron-hole pair recombination, moisture dissolution of perovskite material and photo-oxidation processes at the interface between hole transport materials (HTM) and metal electrode are shown. Adapted with permission from [9]. Copyright 2018 Royal Society of Chemistry.

Encapsulation of organic and perovskite solar cells can play an effective role in improving the stability of both devices. The encapsulation layer can act as a barrier layer by restricting the diffusion of oxygen and moisture through this encapsulation material, resulting in the protection of the cathode interface and the active layer from deterioration as shown in Figure 2. Encapsulation materials should have high barrier performance for oxygen and moisture. The encapsulation material layer structure is a critical factor to overcome these issues and enhance the device stability [10].

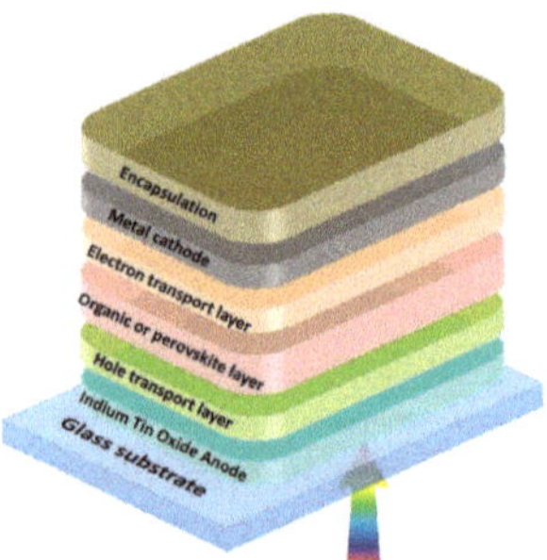

Figure 2. A schematic diagram of an organic or perovskite solar cells structure with an encapsulation layer.

The encapsulation material needs to possess good process ability, low water absorptivity, and permeability. Encapsulation materials should have relatively higher dielectric constant, light transmission, and resistance to ultraviolet (UV) degradation and thermal oxidation. They also require excellent chemical inertness, adhesion, and mechanical strength [11,12]. Oxygen transmission rate (OTR) and water vapour transmission rate (WVTR) are the steady state rates at which oxygen and water vapour gas can penetrate through a film that affects the encapsulation layer. A schematic diagram of organic and/or perovskite solar cells device is shown in Figure 2 with an encapsulation layer. This encapsulation layer material should have the following (Table 1) required properties to protect the device from the oxygen and water vapour effects. A list of specifications of requirements of encapsulation materials is listed in Table 1.

Table 1. Specifications and requirements for encapsulating materials for the protection of organic and perovskite devices from the oxygen and moisture. Reprinted with permission from [12]. Copyright 2016 John Wiley & Sons, Inc.

Characteristics	Specification of Requirement
WVTR	10^{-3}–10^{-6} g·m^{-2}·day^{-1}
OTR	10^{-3}–10^{-5} cm^{3}·m^{-2}·day^{-1}·atm^{-1}
Glass transition temperature (T_g)	<-40 °C (during the winter in deserts)
Total light transmission	>90% of incident light
Water absorption	<0.5 wt % (20 °C/100% RH)
Tensile modulus	<20.7 mPa (>3000 psi) at 25 °C
UV absorption degradation	None (>350 nm)
Hydrolysis	None (80 °C, 100% RH)

To determine the value of OTR, it is necessary to use a colorimetric sensor [13,14]. The OTR was calculated by measuring the amount of oxygen during a certain period at a constant rate that it passed through the cathode. By knowing the value of OTR and WVTR it is possible to determine how good is the encapsulated materials performance for the protection of perovskite and OPVs devices from degradation through their lifetime and performance. The effective WVTR can be determined by monitoring the temporal rate of change of the cathode (metal) electrical conductance. If it is assumed that the diffusivity of water vapour and oxygen obeys Fick's law (diffusivity is independent of concentration), then the WVTR or OTR can be described as [15]:

$$\text{WVTR}(t) = \frac{DC_s}{l}\left[1 + 2\sum_{n=1}^{n}(-1)^n\, e^{\frac{|-Dn^2\pi t|}{l^2}}\right] \tag{1}$$

where D is the diffusivity, C_s is the surface saturation concentrations, t is the time, and l is the film thickness. Normally, D increases with increasing temperature. WVTR and OTR is also a function of temperature.

The good performance encapsulation materials should have a value larger than 10^{-6} g·m^{-2}·day^{-1} for WVTR (Table 1) to protect the OPV and PSCs devices [16]. As an example, an OPV device with a structure ITO/ZnO/P3HT:PCBM/PEDOT:PSS/Al was encapsulated using ZnO and UV resin with a large WVTR value of 5.0 × 10^{-1} g·m^{-2}·day^{-1} [17]. In Ref. [18], the WVTR value was found as big as of 100 g·m^{-2}·day^{-1} for an encapsulated OPV (ITO/PEDOT:PSS/P3HT:PC$_{71}$BM/Ca/Al) with an epoxy resin. On the other hand, in Ref. [17] OPVs cells under the configuration ITO/DMD/Cs$_2$CO$_3$/P3HT:PCBM/MoO$_3$/Al were encapsulated using polyvinyl butyral (PVB), ethylene vinyl acetate (EVA), and thermoplastic poly-urethane (TPU) with a WVTR value of 60, 40, and 150 g·m^{-2}·day^{-1}, respectively, and with an OTR value in the range of 10^{-2}–10^2 cm^{3}·m^{-2}·day^{-1}. In this current work, the WVTR and OTR values were not calculated due to different experimental details.

The most common type of an encapsulation method refers to thin film layers encapsulated on top of OPV and perovskite devices using atomic layer deposition (ALD) [19,20]. ALD is particularly suitable for organic and flexible electronics. However, the ALD technique is expensive. Other methods are roll lamination systems encapsulating the OPV devices between two sheets uniting them with an adhesive [21], the other method is based on heat sealing, a process that basically consists of supplying thermal energy on outside of package to soften/melt the sealants [14] and using a glass substrate that is to be sealed with thermosetting epoxy, it could not be effectively applied to flexible devices [22].

Many works have been done for the advancement of good quality encapsulation technologies to stop the migration of environmental oxygen and moisture into device layers. The device lifetime and stability can be improved with high barrier performance encapsulation materials and structures for providing sufficient durability for commercial application. The encapsulation materials with high optical transmission and high dielectric constant need to possess good processing ability. Mechanical strength, good adhesion, and chemical inertness are also required for a suitable encapsulation material.

It is also expected to have low water-absorptivity and permeability and relatively high resistance to UV degradation and thermal oxidation [23]. Organic materials with a good combination of these properties are most commonly used as the encapsulate for the improvement of acceptable device durability [24]. When compared to their inorganic counterparts, the organic encapsulation materials have demonstrated notable advantages, such as the flexible synthesis of the organic molecules by varying their energy levels, molecular weight, and solubility [25]. Organic encapsulation materials are also expendable and they have lesser impact on the environment [26]. Hence, organic materials are the best suitable candidate for the encapsulation of flexible organic and perovskite devices.

2. Degradation Mechanism of OPV and Perovskite Solar Cells

The degradation of organic and perovskite devices can be divided into two mechanisms: intrinsic and extrinsic. Both of these mechanisms are related to mass-transport processes. The metal electrode/active layer interfaces play major roles in the degradation of both devices [27]. Under normal environmental conditions, the relatively high sensitivity of organic and perovskite materials towards oxygen and moisture decrease the device reliability and lifetime.

2.1. Degradation of OPV Devices

Intrinsic degradation: It is due to changes in the structures of the interfaces between layers of the stacking materials, owing to internal modification of the materials used. The key issue is stability, which is limiting the practical applications of OPV devices. The lifetime of OPV devices is now over many thousands of hours. These improvements of lifetime have been achieved with the application of device architecture optimizations and different interface materials, especially after the investigation of hole conductor layers incorporating carbon electrodes. This is promised stable, low cost, and easy device fabrication methods.

Extrinsic Stability: It is affected by the infiltration of air, e.g., oxygen and water. Such degradation from external factors can escalate by light irradiation. There are some organic materials and metal electrodes can also be degraded by oxygen and water. Oxygen or moisture can be trapped during the fabrication processes or they could diffuse into the cell structure during device lifetime. Due to the following factors, oxygen strongly influences the extrinsic stability in some OPV devices [28]: (i) fullerene molecules are hydrophobic and does not react with water; (ii) the exposure to oxygen in air has negative impact on the electron-transport properties of fullerene; and (iii) oxygen forms surface dipoles and increases the metal work-functions. The abovementioned properties yield in the deterioration of the performance of conventional OPVs; however, they might initially enhance the performance of inverted OPV [11]. The chemical degradation processes are the degradation of the metallic electrodes, degradation of the transparent electrode, intermediate hole extraction layers, and/or even the chosen method to synthesize the materials [29,30]. Although it is well-known that oxygen is typically a p-type dopant in semiconductors, the oxygen vacancies act as electron donors [31]. The electronic properties of semiconducting p-type layers might be improved briefly upon exposure to oxygen [32]. To overcome this issue, researchers have developed inverted OPV devices [33,34]. In a humid environment, the oxide layer can double in thickness, and hence block charge-tunnelling. Unlike the abovementioned positive/negative effects of oxygen, there is hardly any positive impact of water on OPV devices that has ever been suggested. Encapsulation delays the process of degradation, but the currently available materials used for encapsulation. The overall device degradation is not stopped, even by a complex encapsulation schemes, such as a sealed glass container or a high vacuum chamber are employed, because the processes involving water and oxygen are efficiently alleviated. The physical and chemical characteristics of the constituent materials are a complex phenomenon. Several processes may take place simultaneously in both physical and chemical characteristics [35].

2.2. Degradation of Perovskite Devices

There are several issues that are related to the degradation of PSCs, such as interface and device instability. These must be addressed to achieve good reproducibility with high conversion efficiencies and long lifetimes of PSCs. A comprehensive understanding of these issues in PSCs is required to achieve stability breakthroughs for practical commercial applications. For the PCE improvement, the stability of PSCs has to be improved for successful small- and large-scale applications.

The degradation of PSCs can occur as a result of intrinsic and extrinsic stability, such as thermal instability (intrinsic stability) and susceptibility of the perovskite material to ambient conditions (e.g., oxygen and humidity). Degradation can occur by other device structure components, such as the degradation of the ETL (TiO_2 or ZnO) at the interface upon light irradiation and the lack of stability of the hole-transport material [36,37]. The stability of perovskite thin films and PSCs has been widely studied [38–40], which includes the degradation of the perovskite material upon exposure to illumination, humidity, or increased temperature [41–44]. Different schemes have been investigated to enhance the stability of PSCs, such as replacing the mesoporous layer [45], modifying the composition or deposition process of perovskite material [46–49], using various charge-transport materials [50–55] and carbon-based electrodes [46], and interfacial layers and/or surface treatments [56–59]. In most of the researches on stability, the stability tests were reported after the devices being stored in the ambient or dark with measurements of the device performance at regular intervals [52,54,55,60]. The report on encouraging stability of the perovskite devices, such as 0.3% PCE drop in one month, often involves storing them under dark in a low-humidity desiccator [39].

However, it is now well established that PSCs are very susceptible to high humidity. The degradation is notably quicker under high humidity conditions [61,62], and it is much faster under continuous illumination than degradation involving storage in the dark [56,62]. Even the degradation under ambient illumination is significantly faster when compared to dark storage [63]. Since there is no standard testing protocol for reporting stability test results, it is strenuous to draw comparisons directly between test results from different publications. For example, Zhu et al. performed stability tests under constant illumination over a short time period in an N_2 glovebox to prevent the intrusion of humidity [62]. Thermal stability has been reported without exposure to illumination or humidity, or a combination of thermal stress and illumination [64]. In some studies, a white light-emitting diode (LED) was used for the illumination, not a solar simulator [50]. The stability can be over-estimated under illumination, owing to the absence of strong UV component in the emission spectrum. The degradation of the perovskite film is reportedly slower under illumination when a UV filter is used in order to eliminate the UV component from the simulated sunlight illumination [43]. Some works have applied several stress factors at the same time during the tests. The factors include constant illumination, ambient humidity of +50%, and increased temperature [65]. Moreover, the moderately humid environment has been used to obtain outdoor testing results [66].

Multiple stability tests have been performed on cells under low ambient humidity without encapsulation, typically under dark conditions [38,46–49], with very few exceptions [50,57,67]. After several hours, devices without encapsulation generally displayed severe degradation under continuous illumination whilst encapsulated devices showed a relatively longer lifetime [39]. In comparison to fully encapsulated devices with a protected edge, partially encapsulated devices showed a shorter lifespan [46]. This signifies the importance of complete encapsulation in order to enhance extend the lifetime of OPV and perovskite devices. However, the comprehensive comparison between different encapsulation techniques for PSCs is still rare [65]. A careful and complicated encapsulation system, developed by employing the combination of a desiccant material and UV epoxy resin, exhibited significantly improved stability. However, the device performance quickly decreased to <50% of the initial efficiency within 10 h under constant illumination and at an ambient humidity of 80% and cell temperature of 85 °C [65]. The absence of standardized testing conditions and studies regarding the effect of encapsulation on the device stability is hampering the advancements of stable organic and perovskite cells. Various encapsulation methods have been investigated for common perovskite devices that include planar TiO_2 on fluorine doped tin oxide (FTO) glass, methylammonium lead iodide (MAPbI$_3$) as the active layer, and tetrakis[*N,N*-di(4-methoxyphenyl) amino]-9,9′-spirobifluorene (spiro-OMeTAD) as the hole transport layer (HTL) [43]. The international summit on OPV stability (ISOS) was followed for the organic solar cell testing (ISOS-L-2 and ISOS-O-1) [68].

2.2.1. Thermal and Photo Stability

It is found that the photo-stability of perovskite materials is a serious problem for the application of PSCs. The perovskite material begins to decompose into PbI$_3$ at temperature 140 °C [69], and even at temperature 85 °C when it is heated in nitrogen for 24 h [41]. The slight heating of perovskite materials improved the performance of devices due to the formation of a small amount of PbI$_3$ which passivate the perovskite. The device performance may be recovered if it is stored in the dark for a short time. Poor photo-stability of perovskite materials could be raised from the local phase change under an elevated temperature when exposed to light. The dual ion hybridization by compositional material engineering at two sites (A and X) simultaneously can solve the perovskite materials stability. The existing ionic components within the perovskite crystal structure would be partially substituted by analogous ions of similar electronic properties, which would passivate the vulnerable sites present within the structure and thereby eliminating the degradation pathways efficiently from the inside-out. For example—the existing monovalent (A-site) organic cation would be partially substituted by metal (Caesium, Cs$^+$) ions. It has been experimentally proved that heat rather than moisture was the main cause of PSC degradation [70]. The key stability of PSC is to prevent the escape of volatile decomposition products from the perovskite solar cell materials. Polyisobutylene (PIB) encapsulation is one of the promising low cost and low application temperature packaging solutions for PSCs.

2.2.2. Ion Movement

Ion migration in the perovskite materials is another critical issue for the PSCs [71]. The ions (anion or cation) in the halide perovskite materials can move under bias voltage or thermal drift, causing device instability. A schematic diagram of different ion/defects movements in perovskite device structure is shown in Figure 3.

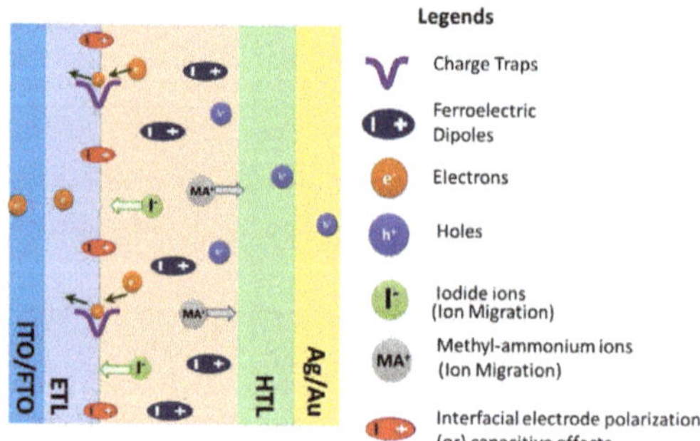

Figure 3. Schematic diagram shows the various ion/defects movements in perovskite device structure that could possibly lead to hysteresis phenomenon during current-voltage characteristic measurements. The exact nature of each ion-movement processes and its degree of influence on hysteresis varies with respect to the device structure and materials involved. Reprinted with permission from [71]. 2018 Solar Energy Materials and Solar Cells.

This defects/ion migration, such as iodine vacancies across the interface, can induce interfacial degradation. This defect migration affects device operational mechanisms and finally cause device failure during operation [9,72].

2.2.3. Electrode Degradation

The interfacial stability of PSCs is also critical for the overall device stability. Recently, it has been found that reactions between iodine-based decomposition products, such as HI from perovskite materials, and traditional metal electrode materials, such as Al or Ag, can lead to the formation of AlI_3 or AgI, respectively [55]. The formation of AlI_3 or AgI compound escalates perovskite decomposition when the material is exposed in an ambient environment (Figure 3) [55]. One way of enhancing the device stability is to insulate the perovskite material from the electrode (Ag/Al). The dispersal of halide ions in PSCs through the charge transport layers could exert a negative impact on the long-term stability of cells.

2.2.4. Charge Transport Layers Degradation

The reasons behind poor device stability of PSCs are due to either the perovskite materials or charge transport layer interfaces. Organic semiconductor layers in the PSC structure that are used as charge transport layers (ETL and HTL) are prone to both oxidization and water absorption that can reduce the device stability. For example, an n-type fullerene, PCBM was used as an efficient ETL in an inverted structure PSC [73]. The authors found that the PCBM went through changes in terms of chemical states or band structure in ambient air, which was a major contributing factor to the device degradation [73]. A p-type organic semiconductor, PEDOT:PSS, is generally used in the inverted structure as the HTL. PEDOT:PSS has high water absorptivity and acidic properties causing etching of the transparent conductive electrode, such as ITO. The hygroscopic and acidic nature of PEDOT:PSS accelerates device degradation [55,74]. Lithium (Li) doped Spiro-OMeTAD salt is also hygroscopic and can diffuse water into the light absorbing perovskite layer causing device failure [75]. Several investigations suggested that oxygen de-absorption from TiO_2 surface (degradation of TiO_2 layer), while light soaking leads to failure of devices [45]. Multiple techniques have been employed for the improvement of the device stability [7,74]. Idigoras et al. [76] deposited thin polymer layer by the remote plasma vacuum deposition of adamantine powder for the encapsulation of perovskite solar cells. They observed that the deterioration of device performance was significantly delayed because of the encapsulation layer.

Bella et al. [77] investigated the stability of PSCs by encapsulation of photochemical resistance as well as moisture tolerance, a fluoro-polymeric light-curable coating on the back contact side. The use of the fluoro-polymeric layer on top of the PSC has the prospect of preventing the oxygen and moisture migration between the top back contact and the perovskite layer due to the hydrophobicity of the coating. The fluoro-polymeric UV-coating possesses a cross-linked nature that can lower the free volume when compared to the traditional non–cross-linked, polymeric systems. Hence, it is expected to improve the long-term stability of PSCs [78]. Figure 4 displays the test results of device stability in terms of PCEs under various atmospheric conditions and photochemical external stresses during 180-days (4320 h). Moreover, those encapsulated devices were tested in outdoor conditions for a time of >3 months (i.e., 2160 h). Successful tolerance toward heavy rain, soil, and dust were observed on the external surface made of glass. In real outdoor conditions, the fluorinated luminescent down shifting layer with low-surface energy helps to clean the front electrode easily.

Uncoated and front-coated devices were studied for an overall period of six months in the aging test. The devices were stored in a glove box filled with Ar for the first three months. The devices were continuously illuminated for 8 h every day with a UV optical fibre with light intensity 5 mW·cm^{-2}. 5% contribution from the UV light source (280–400 nm) helped to simulate the solar spectral irradiance on Earth (AM1.5G, 1000 W·m^{-2}). In the forced UV aging test (Figure 4), the uncoated devices (black curve) had a reduction of 30% in their initial efficiency after one week of UV exposure and stopped working after one month. All of the five front-coated cells (red curve) retained 98% of their initial PCE, even after three months and demonstrated excellent stability under the same conditions.

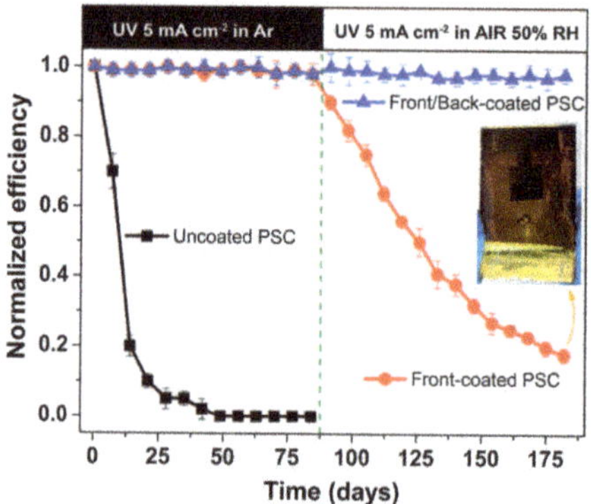

Figure 4. The aging test results on the three series of PSCs: (i) uncoated, (ii) front-coated with a luminescent fluorinated coating, and (iii) front/back-coated (front coating with the luminescent fluorophore and back contact coating with a moisture resistant fluoro-polymeric layer). The cells were stored in an Ar environment during the first three months of the test period. They were kept in the air with RH = 50% for the next three months. During the whole testing period, the PSCs were exposed to continuous UV illumination. The device power conversion efficiency (PCE) was recorded once a week. The inset shows a photograph of a front-coated solar cell at the end of the test. Reprinted with permission from [77]. Copyright 2018 American Association for the Advancement of Science.

By simple visual inspection, the effects of degradation were easily detectable. The progressive yellowing was observed in the mixed-perovskite layer upon 50% RH exposure. The back-contact side was coated with a fluoropolymeric light-curable coating in the front/back-coated devices for stabilizing the devices with enhanced photochemical resistance and moisture tolerance. A similar method was used in the luminescent downshifting experiment for back-coated PSCs. It is hypothesized that the use of highly hydrophobic fluoropolymeric layer on the PSC top can prevent water permeation from the top back contact to the perovskite layer. In addition, the fluoro-polymeric UV-coating possesses a cross-linked nature, leading to lower free volume when compared to the traditional non–cross-linked, polymeric systems. Hence, the coating is expected to improve the long-term stability of PSCs [79]. The aging test results of the front/back-coated PSCs (blue curve), i.e., devices with a luminescent coating on front and a moisture-resistant coating on the back contact, is shown in Figure 4. Even under the combined effects of photochemical and environmental stresses, all the five devices retained exceptional stability of 98%.

The front/back-coated PSCs showed longer lifetime stability retaining most of their initial efficiency the ageing test due to the following reasons:

- securing the perovskite material from UV irradiation and converting it into visible photons;
- shielding devices from oxygen and moisture, hence blocking the hydrolytic behaviour of the perovskite material;
- maintaining clean front electrode clean by the self-cleaning characteristics of this fluorinated polymer. Similar results were also observed for outdoor tests performed.

The non-homogeneous deposition of the coating layers on the front/back side of the device can cause slow hydrolysis of the active perovskite layer. To test the homogeneity of the deposited coating layer, Bella et al. [77] dipped the devices with front/back coatings into the water for about a day. No change in photovoltaic performance was observed in the front/back-coated devices, even after one day of immersion in water.

3. Encapsulation Requirements

To ensure satisfactory encapsulation (Table 1), the permeation of WVTR and OTR needs to be two-three orders of magnitude lower than the bare substrate. Silicon-based dielectric films, deposited by PECVD, have reportedly demonstrated excellent transparent, single-layer barrier performance [80]. However, on polymer substrates it is very difficult to obtain permeation levels that are 1000 times lower by a single layer barrier. The increase in single-layer barrier thickness results in an OTR value that is on-zero asymptotic and is governed by the defects in the barrier layer. Such defects are typically generated from the intrinsic or extrinsic roughness of surface [81]. Hence, the performance of single-layer, encapsulated gas barriers is controlled by the nano-meter size structural defects. In theoretical calculations, the total permeation rate is much higher due to a large number of small pin-holes than that from few but larger defects in the same total pin-hole area [82]. The lateral diffusion of gas is more crucial when the defect diameters are smaller than the substrate thickness. The multilayered deposition of inorganic materials can slightly enhance the performance of the encapsulation barrier. The deposition of organic–inorganic multilayers films to obtain increased barrier performance is the most common practice [83]. However, there are several challenges in the way of utilization and development of thin-film encapsulation. To face these challenges, a variety of materials have been developed to improve the device lifetimes, such as titanium oxide (TiO_x). Using thin films as barriers for device encapsulation is not sufficient for the protection of organic and perovskite devices. The use of glass plates as encapsulation is the simplest and best example for the protection of oxygen and moister. Glass encapsulation can supply the required oxygen and moisture protection in the organic and perovskite solar cells for commercial application. However, glass encapsulation cannot apply on inflexible devices. It has also been shown that a thin layer of SiO_xC_y acts as a barrier for oxygen and moisture. Unfortunately, the PECVD process is required for the deposition of this encapsulation film, which is expensive and uses complicated vacuum systems. This is also not suitable to deposit on a flexible substrate as a barrier of oxygen and moister transmission. Although there are other encapsulating materials, they cannot fulfil the barrier requirements for organic and perovskite solar cells. In this review paper, we have presented different encapsulation materials used in various approaches and integrated with organic and perovskite devices. In Table 2, we have summarized some of the polymer materials with their WVTR values for the organic and perovskite solar cells. Recently, using the RF plasma polymerization technique linalyl acetate [84,85] and terpinene-4-ol thin films has been deposited for the encapsulation of devices. On the basis of surface and optical properties, these polymer based thin films can be potential encapsulating layers for organic and perovskite solar cells [86,87].

Materials for Encapsulation

The encapsulation materials should have high a dielectric breakdown that matches the refractive index with other layers and high volume resistivity. Materials should be low cost, dimensionally stability, and easy to deposit. WVTR and OTR represent the steady state rates at which water vapour and oxygen gas, respectively, penetrate through the encapsulating film that affects the encapsulation layer. Glass transition temperature (T_g) for organic encapsulation material is another important property, which is dependent on the chain flexibility. The polymer mechanically varies from being rigid and brittle and becomes tough and leathery. The maximum exposure temperatures on encapsulation material and the effect on the mechanical behaviour of the material should be known. The light transmission through the encapsulation materials is also important to measure to understand how it will affect the device performance. The encapsulation material requirements are also to define the UV absorption degradation, hydrolysis, and some other aspects. A list of suitable polymer encapsulation materials and their specifications for the roll-to-roll fabrication of devices are given below.

Table 2. List of probable polymer encapsulation materials for organic photovoltaic (OPV) and Perovskite devices.

Materials	Encapsulation Type	Water Vapour Transmission Rate (WVTR) ($g \cdot m^{-2} \cdot day^{-1}$)	Comments	References
Ethylene vinyl acetate (EVA)	Single layer encapsulation	40	Light transmission of 91%. Suitable for resisting weather and long-term reliability under light exposure. It is suitable for encapsulation of organic and perovskite solar modules.	[88,89]
Europium doped EVA: Eu^{3+}	Single layer encapsulation	40	Absorption bang gap is 310 nm (4 eV). Suitable for PV module encapsulation. Eu^{3+} doped EVA layers can induce photon down-shifting with wavelengths <460 nm.	[90]
Ethylene methyl acrylate (EMA)	Single layer encapsulation	Not mentioned	EMA is suitable for chemical resistance, thermally stability, adherence to different substrates and excellent mechanical behaviour at low temperature. It is suitable for encapsulation of perovskite and organic devices.	[23,91]
Polyvinyl butyral (PVB)	Single layer encapsulation	60	PVB has high optical transparency, good heat resistance, good adherence to solar cells, glass, and other plastics, increased bond durability, and compatible with module components. PVB is already used as encapsulation layer for thin film solar cells.	[14]
Thermoplastic polyurethane (TPU)	Single layer encapsulation	150	TPU film is better than EVA film for encapsulation since it is flexible to bond with relatively hard materials. These films can be processed in normal atmospheric pressure without cross-linking and emissions.	[92,93]
UV-cured epoxy	Single layer encapsulation	16	Epoxy film is good for encapsulation. It is optically transparent, thermally conductive, weather resistance, high temperature resistance, good adhesive properties on glass and plastic.	[70]
Polyisobutylene (PIB)	Single layer encapsulation	0.001–0.0001	PIB is a synthetic rubber. It can encapsulate organic and perovskite solar cells.	[70]
Cyclized perfluoro-polymer (CytopTM)	Single layer encapsulation	Not mentioned	Conventional thin-film deposition techniques e.g., spin coating can be used to deposit this polymer. It is transparent and amorphous. This material is good for weather resistant, good for oxygen/water-vapor shielding for testing organic device lifetime.	[94]
Organic–inorganic hybrid materials ORMOCERs (ORM)	Single layer encapsulation	0.01	It is organic and inorganic components modified ceramics or ORM. It has good chemical resistance, highly transparent, anti-soiling, diffusion- inhibition. OTR is <0.01 cm^3 m^{-2} day^{-1}. These properties are necessary for the encapsulation of organic and perovskite solar cells.	[95,96]
ORMOSIL aero-gel thin film	Single layer encapsulation	Not mentioned	It is mechanically and thermally stable and highly transparent. ORMOSIL materials are flexibility and stability at atmospheric conditions. ORMOSIL has variable organic group that can modify the chemical and physical surface properties.	[97,98]
Other organic materials	Single layer encapsulation	Not mentioned	Encapsulation tested of 10 polymers were poly(vinyl methyl ketone) (PVMK), poly(methyl methacrylate) (PMMA), poly(vinylidene chloride-co-vinyl chloride) (PVDC–co-PVC), poly (vinylidene fluoride)(PVDF), polyacrylonitride (PAN), poly(vinylalcohol) (PVA), poly(vinylphenol) (PVP), poly(methyl vinyl ether) (PMVE), polystyrene (PS), and poly(vinyl chloride) (PVC). An encapsulation approach with these 10 low-polarity polymer is demonstrated to block water/moisture and prevent encapsulation-induced degradation.	[99,100]
Luminescent downshifting fluoro-polymeric coating such as PTFE (polytetrafluoroethylene), PFA (perfluoroalkoxy), FEP (fluorinated ethylene-propylene), etc.	Single layer encapsulation	Not mentioned	It is very good as an encapsulating material for organic and perovskite solar cells to improve the device stability for the out-door application. Fluoropolymers are excellent for chemical and thermal resistance. Their surfaces are non-reactive with all chemicals and solvents.	[101]

4. Discussion

Thin-film encapsulation is a vital technology that is required for the application and commercialization of perovskite and organic solar cells. An effective encapsulation is crucial to prevent the permeation of moister and oxygen to achieve the desired reliability and device lifetime. More progress is needed to develop encapsulating materials for devices with specific requirements. The processing conditions of EVA exposure time and damp heat affect its adhesion strength. Transport of moister and oxygen through local pinholes become an issue for achieving ultrahigh barriers encapsulation that is hard to be avoided. New encapsulation materials need to be developed to fulfil the encapsulation requirements and achieve ultrahigh encapsulation barriers. Furthermore, the encapsulation processing temperature of thin film should be within the temperature range that is suitable for the organic polymer substrate materials. Usually, low-temperature PECVD processed inorganic encapsulation thin films are not suitable for the encapsulation barrier, since these films suffer from intrinsic defects. Water vapour and oxygen can easily diffuse through these defects. For high performance encapsulation barriers, these intrinsic defects should be reduced or passivated. Inorganic and organic alternating multilayer films can be one solution to reduce and avoid intrinsic pinholes. The defects in the inorganic layers can be passivated by an organic film and do not channel continuously through the multilayer structure. A steady state permeation rate of moister and oxygen can be maintained by the multilayer encapsulation barrier structures. A transient rate of encapsulation barrier can be maintained over a specific time-period that might exceed the desired lifetime of encapsulated devices. Typically, the permeation rate in the transient region is lower than the permeation rate in the steady state. It is suggested that the barrier performance characterization from the initial transient period will provide an underestimation of the total permeation rate for long-term applications. As a result, the barrier performance in multilayer structures should be characterized separately, such as the steady-state and transient permeation rates. This should be used for the calculation of lag time to avoid overestimation of expected device lifetimes or barrier performance during both storage and application. Moreover, the deposition of high quality inorganic encapsulation layer with vacuum deposition processes is expensive and low through put.

Since there are alternative deposition processes using different deposition chambers for organic and inorganic layers, multilayer barrier structures are expensive. If inorganic layers can be deposited from the solution precursors, it can be a promising solution to this problem, making it possible to deposit inorganic and organic multilayer barriers employing similar deposition techniques. Hence, totally solution process should be developed for low cost, high barrier materials for perovskite and OPV encapsulation. Optimization of the individual film dimension can be useful multilayer encapsulation structures. Before adapting these technologies, the quality of films should also be improved. The reliability and continuous yield of encapsulation processes must be explored to reduce the processing steps and time. Since perovskite and organic solar cells are very susceptible to moisture, it is crucial to use a low-diffusivity edge seal material.

Graphene is a carbon-based one-dimensional material with excellent electronic and mechanical properties offering limitless opportunities in device engineering. In line with this, graphene can be used as a bi-functional electrode that will serve as a highly conductive charge collection electrode as well as an encapsulant for the underlying organic or perovskite layers. This carbon-graphene based encapsulation technology is sought to be a viable approach for large scale commercialization and deployment of organic and perovskite solar cells technology.

5. Conclusions

Effective thin film encapsulation is crucial to prevent the permeation of water vapour and oxygen for achieving the stability and desired life times of organic and perovskite solar cells. The problem of achieving a thin layer encapsulation barrier is transport-dependent permeation through localized pin-holes that is strenuous to control. New materials and technology are required to satisfy the requirements of encapsulation to prevent the permeation of moisture and oxygen for the reliability and stability of devices. The encapsulation processing temperature of thin film should be within the temperature range suitable for the organic and polymer substrate materials. Multilayer polymer films of encapsulation can be a solution to block the pinholes in the films to stop the permeation of water vapour and oxygen. However, the barrier performance of multilayer structures encapsulation consists of a steady state permeation rate as well as a transient rate over a specified time. These rates of encapsulation barrier can exceed the desired lifetime of encapsulated devices. The barrier performance in multilayer structures should be characterized separately by the steady-state and transient permeation rates over a specified time-period. Multilayer thin film encapsulation structures deposited under vacuum will be exorbitant, owing to alternative deposition of layers in separate deposition chambers. Solution process deposition can be a promising alternative for low-cost barrier materials for thin film multilayers encapsulation. Subsequently, multilayers can be deposited with a similar deposition process to obtain low-cost barrier materials. Film quality should be improved before adopting the multilayer encapsulation technologies. Encapsulation layers with high quality are significant to improve the device lifetimes.

Funding: This research received no external funding.

Acknowledgments: We acknowledge the financial support from Future Solar Technologies Pty. Ltd. for this work. We also acknowledge the endless support from the staff of photovoltaic and renewable energy engineering school (SPREE), UNSW, Sydney.

Conflicts of Interest: The authors declare no conflict of interest. The funders had no role in the design of the study; in the collection, analyses, or interpretation of data; in the writing of the manuscript, and in the decision to publish the results.

References

1. Liang, P.-W.; Liao, C.-Y.; Chueh, C.-C.; Zuo, F.; Williams, S.T.; Xin, X.-K.; Lin, J.; Jen, A.K.-Y. Additive Enhanced crystallization of solution-processed perovskite for highly efficient planar-heterojunction solar cells. *Adv. Mater.* **2014**, *26*, 3748–3754. [CrossRef] [PubMed]

2. Li, H.; Xiao, Z.; Ding, L.; Wang, J. Thermostable single-junction organic solar cells with a power conversion efficiency of 14.62%. *Sci. Bull.* **2018**, *63*, 340–342. [CrossRef]

3. Meng, L.; Zhang, Y.; Wan, X.; Li, C.; Zhang, X.; Wang, Y.; Ke, X.; Xiao, Z.; Ding, L.; Xia, R.; et al. Organic and solution-processed tandem solar cells with 17.3% efficiency. *Science* **2018**, *361*, 1094–1098. [CrossRef] [PubMed]

4. Xiao, Z.; Jia, X.; Ding, L. Ternary organic solar cells offer 14% power conversion efficiency. *Sci. Bull.* **2017**, *62*, 1562–1564. [CrossRef]

5. Xu, C.; Wright, M.; Ping, D.; Yi, H.; Zhang, X.; Mahmud, M.A.; Sun, K.; Upama, M.B.; Haque, F.; Uddin, A. Ternary blend organic solar cells with a non-fullerene acceptor as a third component to synergistically improve the efficiency. *Org. Electron.* **2018**, *62*, 261–268. [CrossRef]

6. Green, M.A.; Hishikawa, Y.; Dunlop, E.D.; Levi, D.H.; Hohl-Ebinger, J.; Ho-Baillie, A.W. Solar cell efficiency tables (version 51). *Prog. Photovolt. Res. Appl.* **2018**, *26*, 3–12. [CrossRef]

7. Mahmud, M.A.; Elumalai, N.K.; Pal, B.; Jose, R.; Upama, M.B.; Wang, D.; Goncales, V.R.; Xu, C.; Haque, F.; Uddin, A. Electrospun 3D composite nano-flowers for high performance triple-cation perovskite solar cells. *Electrochim. Acta* **2018**, *289*, 459–473. [CrossRef]

8. Wang, D.; Wright, M.; Elumalai, N.K.; Uddin, A. Stability of perovskite solar cells. *Sol. Energy Mater. Sol. Cells* **2016**, *147*, 255–275. [CrossRef]

9. Berhe, T.A.; Su, W.-N.; Chen, C.-H.; Pan, C.-J.; Cheng, J.-H.; Chen, H.-M.; Tsai, M.-C.; Chen, L.-Y.; Dubale, A.A.; Hwang, B.-J. Organometal halide perovskite solar cells: Degradation and stability. *Energy Environ. Sci.* **2016**, *9*, 323–356. [CrossRef]

10. Li, F.; Liu, M. Recent efficient strategies for improving the moisture stability of perovskite solar cells. *J. Mater. Chem. A* **2017**, *5*, 15447–15459. [CrossRef]

11. Ahmad, J.; Bazaka, K.; Anderson, L.J.; White, R.D.; Jacob, M.V. Materials and methods for encapsulation of OPV: A review. *Renew. Sustain. Energy Rev.* **2013**, *27*, 104–117. [CrossRef]

12. Kang, H.; Kim, G.; Kim, J.; Kwon, S.; Kim, H.; Lee, K. Bulk-heterojunction organic solar cells: Five core technologies for their commercialization. *Adv. Mater.* **2016**, *28*, 7821–7861. [CrossRef]

13. Cros, S.; De Bettignies, R.; Berson, S.; Bailly, S.; Maisse, P.; Lemaitre, N.; Guillerez, S. Definition of encapsulation barrier requirements: A method applied to organic solar cells. *Sol. Energy Mater. Sol. Cells* **2011**, *95*, S65–S69. [CrossRef]

14. Kim, N.; Potscavage , W.J., Jr.; Sundaramoothi, A.; Henderson, C.; Kippelen, B.; Graham, S. A correlation study between barrier film performance and shelf lifetime of encapsulated organic solar cells. *Sol. Energy Mater. Sol. Cells* **2012**, *101*, 140–146. [CrossRef]

15. Crank, J. *The Mathematics of Diffusion*; Oxford University Press: Oxford, UK, 1979.

16. Kippelen, B.; Brédas, J.-L. Organic photovoltaics. *Energy Environ. Sci.* **2009**, *2*, 251–261. [CrossRef]

17. Hauch, J.A.; Schilinsky, P.; Choulis, S.A.; Rajoelson, S.; Brabec, C.J. The impact of water vapor transmission rate on the lifetime of flexible polymer solar cells. *Appl. Phys. Lett.* **2008**, *93*, 103306. [CrossRef]

18. Lee, H.-J.; Kim, H.-P.; Kim, H.-M.; Youn, J.-H.; Nam, D.-H.; Lee, Y.-G.; Lee, J.-G.; bin Mohd Yusoff, A.R.; Jang, J. Solution processed encapsulation for organic photovoltaics. *Sol. Energy Mater. Sol. Cells* **2013**, *111*, 97–101. [CrossRef]

19. Kim, Y.; Kim, H.; Graham, S.; Dyer, A.; Reynolds, J.R. Durable polyisobutylene edge sealants for organic electronics and electrochemical devices. *Sol. Energy Mater. Sol. Cells* **2012**, *100*, 120–125. [CrossRef]

20. Zardetto, V.; Williams, B.; Perrotta, A.; Di Giacomo, F.; Verheijen, M.; Andriessen, R.; Kessels, W.; Creatore, M. Atomic layer deposition for perovskite solar cells: Research status, opportunities and challenges. *Sustain. Energy Fuels* **2017**, *1*, 30–55. [CrossRef]

21. Tanenbaum, D.M.; Dam, H.F.; Rösch, R.; Jørgensen, M.; Hoppe, H.; Krebs, F.C. Edge sealing for low cost stability enhancement of roll-to-roll processed flexible polymer solar cell modules. *Sol. Energy Mater. Sol. Cells* **2012**, *97*, 157–163. [CrossRef]

22. Elkington, D.; Cooling, N.; Zhou, X.; Belcher, W.; Dastoor, P. Single-step annealing and encapsulation for organic photovoltaics using an exothermically-setting encapsulant material. *Sol. Energy Mater. Sol. Cells* **2014**, *124*, 75–78. [CrossRef]

23. Cuddihy, E.; Coulbert, C.; Gupta, A.; Liang, R. *Electricity from Photovoltaic Solar Cells: Flat-Plate Solar Array Project Final Report*; Volume VII: Module Encapsulation; JPL Publication: Pasadena, CA, USA, 1986.

24. Kim, N. *Fabrication and Characterization of Thin-Film Encapsulation for Organic Electronics*; Georgia Institute of Technology: Atlanta, GA, USA, 2009.

25. Spanggaard, H.; Krebs, F.C. A brief history of the development of organic and polymeric photovoltaics. *Sol. Energy Mater. Sol. Cells* **2004**, *83*, 125–146. [CrossRef]

26. Shaheen, S.E.; Ginley, D.S.; Jabbour, G.E. Organic-based photovoltaics: Toward low-cost power generation. *MRS Bull.* **2005**, *30*, 10–19. [CrossRef]

27. Nguyen, T.-P.; Renaud, C.; Reisdorffer, F.; Wang, L.-J. Degradation of phenyl C61 butyric acid methyl ester: Poly (3-hexylthiophene) organic photovoltaic cells and structure changes as determined by defect investigations. *J. Photonics Energy* **2012**, *2*, 021013. [CrossRef]

28. Reese, M.O.; Morfa, A.J.; White, M.S.; Kopidakis, N.; Shaheen, S.E.; Rumbles, G.; Ginley, D.S. Pathways for the degradation of organic photovoltaic P3HT:PCBM based devices. *Sol. Energy Mater. Sol. Cells* **2008**, *92*, 746–752. [CrossRef]

29. Dennler, G.; Lungenschmied, C.; Neugebauer, H.; Sariciftci, N.S.; Latreche, M.; Czeremuszkin, G.; Wertheimer, M.R. A new encapsulation solution for flexible organic solar cells. *Thin Solid Films* **2006**, *511*, 349–353. [CrossRef]

30. Ray, B.; Alam, M.A. A compact physical model for morphology induced intrinsic degradation of organic bulk heterojunction solar cell. *Appl. Phys. Lett.* **2011**, *99*, 140. [CrossRef]

31. Karlsson, K.; Troncale, V.; Oberli, D.; Malko, A.; Pelucchi, E.; Rudra, A.; Kapon, E. Optical polarization anisotropy and hole states in pyramidal quantum dots. *Appl. Phys. Lett.* **2006**, *89*, 251113. [CrossRef]

32. Cao, H.; He, W.; Mao, Y.; Lin, X.; Ishikawa, K.; Dickerson, J.H.; Hess, W.P. Recent progress in degradation and stabilization of organic solar cells. *J. Power Sources* **2014**, *264*, 168–183. [CrossRef]

33. Lin, R.; Miwa, M.; Wright, M.; Uddin, A. Optimisation of the sol-gel derived ZnO buffer layer for inverted structure bulk heterojunction organic solar cells using a low band gap polymer. *Thin Solid Films* **2014**, *566*, 99–107. [CrossRef]

34. Lin, R.; Wright, M.; Chan, K.H.; Puthen-Veettil, B.; Sheng, R.; Wen, X.; Uddin, A. Performance improvement of low bandgap polymer bulk heterojunction solar cells by incorporating P3HT. *Org. Electron.* **2014**, *15*, 2837–2846. [CrossRef]

35. Grossiord, N.; Kroon, J.M.; Andriessen, R.; Blom, P.W. Degradation mechanisms in organic photovoltaic devices. *Org. Electron.* **2012**, *13*, 432–456. [CrossRef]

36. Tiep, N.H.; Ku, Z.; Fan, H.J. Recent advances in improving the stability of perovskite solar cells. *Adv. Energy Mater.* **2016**, *6*, 1501420. [CrossRef]

37. Ye, M.; Hong, X.; Zhang, F.; Liu, X. Recent advancements in perovskite solar cells: Flexibility, stability and large scale. *J. Mater. Chem. A* **2016**, *4*, 6755–6771. [CrossRef]

38. Bi, D.; Gao, P.; Scopelliti, R.; Oveisi, E.; Luo, J.; Grätzel, M.; Hagfeldt, A.; Nazeeruddin, M.K. High-performance perovskite solar cells with enhanced environmental stability based on amphiphile-modified $CH_3NH_3PbI_3$. *Adv. Mater.* **2016**, *28*, 2910–2915. [CrossRef] [PubMed]

39. Bi, D.; Tress, W.; Dar, M.I.; Gao, P.; Luo, J.; Renevier, C.; Schenk, K.; Abate, A.; Giordano, F.; Baena, J.-P.C. Efficient luminescent solar cells based on tailored mixed-cation perovskites. *Sci. Adv.* **2016**, *2*, e1501170. [CrossRef] [PubMed]

40. Shi, Z.; Jayatissa, A. Perovskites-based solar cells: A review of recent progress, materials and processing methods. *Materials* **2018**, *11*, 729. [CrossRef] [PubMed]

41. Conings, B.; Drijkoningen, J.; Gauquelin, N.; Babayigit, A.; D'Haen, J.; D'Olieslaeger, L.; Ethirajan, A.; Verbeeck, J.; Manca, J.; Mosconi, E. Intrinsic thermal instability of methylammonium lead trihalide perovskite. *Adv. Energy Mater.* **2015**, *5*, 1500477. [CrossRef]

42. Li, Y.; Xu, X.; Wang, C.; Wang, C.; Xie, F.; Yang, J.; Gao, Y. Degradation by exposure of coevaporated $CH_3NH_3PbI_3$ thin films. *J. Phys. Chem. C* **2015**, *119*, 23996–24002. [CrossRef]

43. Liu, F.; Dong, Q.; Wong, M.K.; Djurišić, A.B.; Ng, A.; Ren, Z.; Shen, Q.; Surya, C.; Chan, W.K.; Wang, J. Is excess PbI_2 beneficial for perovskite solar cell performance? *Adv. Energy Mater.* **2016**, *6*, 1502206. [CrossRef]

44. Niu, G.; Li, W.; Meng, F.; Wang, L.; Dong, H.; Qiu, Y. Study on the stability of $CH_3NH_3PbI_3$ films and the effect of post-modification by aluminum oxide in all-solid-state hybrid solar cells. *J. Mater. Chem. A* **2014**, *2*, 705–710. [CrossRef]

45. Leijtens, T.; Eperon, G.E.; Pathak, S.; Abate, A.; Lee, M.M.; Snaith, H.J. Overcoming ultraviolet light instability of sensitized TiO_2 with meso-superstructured organometal tri-halide perovskite solar cells. *Nat. Commun.* **2013**, *4*, 2885. [CrossRef] [PubMed]

46. Li, B.; Li, Y.; Zheng, C.; Gao, D.; Huang, W. Advancements in the stability of perovskite solar cells: Degradation mechanisms and improvement approaches. *RSC Adv.* **2016**, *6*, 38079–38091. [CrossRef]

47. Mahmud, M.A.; Elumalai, N.K.; Upama, M.B.; Wang, D.; Wright, M.; Chan, K.H.; Xu, C.; Haque, F.; Uddin, A. Single vs mixed organic cation for low temperature processed perovskite solar cells. *Electrochim. Acta* **2016**, *222*, 1510–1521. [CrossRef]

48. Xia, X.; Wu, W.; Li, H.; Zheng, B.; Xue, Y.; Xu, J.; Zhang, D.; Gao, C.; Liu, X. Spray reaction prepared $FA_{1-x}Cs_xPbI_3$ solid solution as a light harvester for perovskite solar cells with improved humidity stability. *RSC Adv.* **2016**, *6*, 14792–14798. [CrossRef]

49. Zhu, W.; Bao, C.; Li, F.; Yu, T.; Gao, H.; Yi, Y.; Yang, J.; Fu, G.; Zhou, X.; Zou, Z. A halide exchange engineering for $CH_3NH_3PbI_{3-x}Br_x$ perovskite solar cells with high performance and stability. *Nano Energy* **2016**, *19*, 17–26. [CrossRef]

50. Agresti, A.; Pescetelli, S.; Cinà, L.; Konios, D.; Kakavelakis, G.; Kymakis, E.; Carlo, A.D. Efficiency and stability enhancement in perovskite solar cells by inserting lithium—Neutralized graphene oxide as electron transporting layer. *Adv. Funct. Mater.* **2016**, *26*, 2686–2694. [CrossRef]

51. Arafat Mahmud, M.; Kumar Elumalai, N.; Baishakhi Upama, M.; Wang, D.; Haque, F.; Wright, M.; Howe Chan, K.; Xu, C.; Uddin, A. Enhanced stability of low temperature processed perovskite solar cells via augmented polaronic intensity of hole transporting layer. *Phys. Status Solidi (RRL)–Rapid Res. Lett.* **2016**, *10*, 882–889. [CrossRef]

52. Lee, D.-Y.; Na, S.-I.; Kim, S.-S. Graphene oxide/PEDOT:PSS composite hole transport layer for efficient and stable planar heterojunction perovskite solar cells. *Nanoscale* **2016**, *8*, 1513–1522. [CrossRef]

53. Wang, Y.; Yuan, Z.; Shi, G.; Li, Y.; Li, Q.; Hui, F.; Sun, B.; Jiang, Z.; Liao, L. Dopant-free spiro-triphenylamine/fluorene as hole—Transporting material for perovskite solar cells with enhanced efficiency and stability. *Adv. Funct. Mater.* **2016**, *26*, 1375–1381. [CrossRef]

54. Xu, J.; Voznyy, O.; Comin, R.; Gong, X.; Walters, G.; Liu, M.; Kanjanaboos, P.; Lan, X.; Sargent, E.H. Crosslinked remote-doped hole-extracting contacts enhance stability under accelerated lifetime testing in perovskite solar cells. *Adv. Mater.* **2016**, *28*, 2807–2815. [CrossRef] [PubMed]

55. You, J.; Meng, L.; Song, T.-B.; Guo, T.-F.; Yang, Y.M.; Chang, W.-H.; Hong, Z.; Chen, H.; Zhou, H.; Chen, Q. Improved air stability of perovskite solar cells via solution-processed metal oxide transport layers. *Nat. Nanotechnol.* **2016**, *11*, 75. [CrossRef] [PubMed]

56. Chang, C.-Y.; Chang, Y.-C.; Huang, W.-K.; Liao, W.-C.; Wang, H.; Yeh, C.; Tsai, B.-C.; Huang, Y.-C.; Tsao, C.-S. Achieving high efficiency and improved stability in large-area ITO-free perovskite solar cells with thiol-functionalized self-assembled monolayers. *J. Mater. Chem. A* **2016**, *4*, 7903–7913. [CrossRef]

57. Igbari, F.; Li, M.; Hu, Y.; Wang, Z.-K.; Liao, L.-S. A room-temperature $CuAlO_2$ hole interfacial layer for efficient and stable planar perovskite solar cells. *J. Mater. Chem. A* **2016**, *4*, 1326–1335. [CrossRef]

58. Mahmud, M.A.; Elumalai, N.K.; Upama, M.B.; Wang, D.; Gonçales, V.R.; Wright, M.; Gooding, J.J.; Haque, F.; Xu, C.; Uddin, A. Cesium compounds as interface modifiers for stable and efficient perovskite solar cells. *Sol. Energy Mater. Sol. Cells* **2018**, *174*, 172–186. [CrossRef]

59. Song, D.; Wei, D.; Cui, P.; Li, M.; Duan, Z.; Wang, T.; Ji, J.; Li, Y.; Mbengue, J.M.; Li, Y. Dual function interfacial layer for highly efficient and stable lead halide perovskite solar cells. *J. Mater. Chem. A* **2016**, *4*, 6091–6097. [CrossRef]

60. Kim, H.-S.; Lee, C.-R.; Im, J.-H.; Lee, K.-B.; Moehl, T.; Marchioro, A.; Moon, S.-J.; Humphry-Baker, R.; Yum, J.-H.; Moser, J.E. Lead iodide perovskite sensitized all-solid-state submicron thin film mesoscopic solar cell with efficiency exceeding 9%. *Sci. Rep.* **2012**, *2*, 591. [CrossRef] [PubMed]

61. Christians, J.A.; Miranda Herrera, P.A.; Kamat, P.V. Transformation of the excited state and photovoltaic efficiency of $CH_3NH_3PbI_3$ perovskite upon controlled exposure to humidified air. *J. Am. Chem. Soc.* **2015**, *137*, 1530–1538. [CrossRef] [PubMed]

62. Zhu, Z.; Bai, Y.; Liu, X.; Chueh, C.C.; Yang, S.; Jen, A.K.Y. Enhanced efficiency and stability of inverted perovskite solar cells using highly crystalline SnO_2 nanocrystals as the robust electron—Transporting layer. *Adv. Mater.* **2016**, *28*, 6478–6484. [CrossRef]

63. Aldibaja, F.K.; Badia, L.; Mas-Marzá, E.; Sánchez, R.S.; Barea, E.M.; Mora-Sero, I. Effect of different lead precursors on perovskite solar cell performance and stability. *J. Mater. Chem. A* **2015**, *3*, 9194–9200. [CrossRef]

64. Dkhissi, Y.; Weerasinghe, H.; Meyer, S.; Benesperi, I.; Bach, U.; Spiccia, L.; Caruso, R.A.; Cheng, Y.-B. Parameters responsible for the degradation of $CH_3NH_3PbI_3$-based solar cells on polymer substrates. *Nano Energy* **2016**, *22*, 211–222. [CrossRef]

65. Han, Y.; Meyer, S.; Dkhissi, Y.; Weber, K.; Pringle, J.M.; Bach, U.; Spiccia, L.; Cheng, Y.-B. Degradation observations of encapsulated planar $CH_3NH_3PbI_3$ perovskite solar cells at high temperatures and humidity. *J. Mater. Chem. A* **2015**, *3*, 8139–8147. [CrossRef]

66. Li, X.; Tschumi, M.; Han, H.; Babkair, S.S.; Alzubaydi, R.A.; Ansari, A.A.; Habib, S.S.; Nazeeruddin, M.K.; Zakeeruddin, S.M.; Grätzel, M. Outdoor performance and stability under elevated temperatures and long-term light soaking of triple-layer mesoporous perovskite photovoltaics. *Energy Technol.* **2015**, *3*, 551–555. [CrossRef]

67. Xie, F.X.; Zhang, D.; Su, H.; Ren, X.; Wong, K.S.; Graätzel, M.; Choy, W.C. Vacuum-assisted thermal annealing of $CH_3NH_3PbI_3$ for highly stable and efficient perovskite solar cells. *ACS Nano* **2015**, *9*, 639–646. [CrossRef]

68. Reese, M.O.; Gevorgyan, S.A.; Jørgensen, M.; Bundgaard, E.; Kurtz, S.R.; Ginley, D.S.; Olson, D.C.; Lloyd, M.T.; Morvillo, P.; Katz, E.A. Consensus stability testing protocols for organic photovoltaic materials and devices. *Sol. Energy Mater. Sol. Cells* **2011**, *95*, 1253–1267. [CrossRef]

69. Supasai, T.; Rujisamphan, N.; Ullrich, K.; Chemseddine, A.; Dittrich, T. Formation of a passivating CH3NH3PbI3/PbI2 interface during moderate heating of $CH_3NH_3PbI_3$ layers. *Appl. Phys. Lett.* **2013**, *103*, 183906. [CrossRef]

70. Shi, L.; Young, T.L.; Kim, J.; Sheng, Y.; Wang, L.; Chen, Y.; Feng, Z.; Keevers, M.J.; Hao, X.; Verlinden, P.J. Accelerated lifetime testing of organic–inorganic perovskite solar cells encapsulated by polyisobutylene. *ACS Appl. Mater. Interfaces* **2017**, *9*, 25073–25081. [CrossRef]

71. Elumalai, N.K.; Uddin, A. Hysteresis in organic-inorganic hybrid perovskite solar cells. *Sol. Energy Mater. Sol. Cells* **2016**, *157*, 476–509. [CrossRef]

72. Azpiroz, J.M.; Mosconi, E.; Bisquert, J.; De Angelis, F. Defect migration in methylammonium lead iodide and its role in perovskite solar cell operation. *Energy Environ. Sci.* **2015**, *8*, 2118–2127. [CrossRef]

73. Upama, M.B.; Elumalai, N.K.; Mahmud, M.A.; Wang, D.; Haque, F.; Gonçales, V.R.; Gooding, J.J.; Wright, M.; Xu, C.; Uddin, A. Role of fullerene electron transport layer on the morphology and optoelectronic properties of perovskite solar cells. *Org. Electron.* **2017**, *50*, 279–289. [CrossRef]

74. Kim, J.H.; Liang, P.W.; Williams, S.T.; Cho, N.; Chueh, C.C.; Glaz, M.S.; Ginger, D.S.; Jen, A.K.Y. High-performance and environmentally stable planar heterojunction perovskite solar cells based on a solution-processed copper-doped nickel oxide hole-transporting layer. *Adv. Mater.* **2015**, *27*, 695–701. [CrossRef] [PubMed]

75. Ono, L.K.; Raga, S.R.; Remeika, M.; Winchester, A.J.; Gabe, A.; Qi, Y. Pinhole-free hole transport layers significantly improve the stability of MAPbI$_3$-based perovskite solar cells under operating conditions. *J. Mater. Chem. A* **2015**, *3*, 15451–15456. [CrossRef]

76. Idígoras, J.; Aparicio, F.J.; Contreras-Bernal, L.; Ramos-Terrón, S.; Alcaire, M.; Sánchez-Valencia, J.R.; Borras, A.; Barranco, Á.; Anta, J.A. Enhancing moisture and water resistance in perovskite solar cells by encapsulation with ultrathin plasma polymers. *ACS Appl. Mater. Interfaces* **2018**, *10*, 11587–11594. [CrossRef] [PubMed]

77. Bella, F.; Griffini, G.; Correa-Baena, J.-P.; Saracco, G.; Grätzel, M.; Hagfeldt, A.; Turri, S.; Gerbaldi, C. Improving efficiency and stability of perovskite solar cells with photocurable fluoropolymers. *Science* **2016**, *354*, 203–206. [CrossRef] [PubMed]

78. Griffini, G.; Levi, M.; Turri, S. Novel crosslinked host matrices based on fluorinated polymers for long-term durability in thin-film luminescent solar concentrators. *Sol. Energy Mater. Sol. Cells* **2013**, *118*, 36–42. [CrossRef]

79. Griffini, G.; Levi, M.; Turri, S. Novel high-durability luminescent solar concentrators based on fluoropolymer coatings. *Prog. Organ. Coat.* **2014**, *77*, 528–536. [CrossRef]

80. Da Silva Sobrinho, A.; Latreche, M.; Czeremuszkin, G.; Klemberg-Sapieha, J.; Wertheimer, M. Transparent barrier coatings on polyethylene terephthalate by single-and dual-frequency plasma-enhanced chemical vapor deposition. *J. Vac. Sci. Technol. A Vac. Surf. Films* **1998**, *16*, 3190–3198. [CrossRef]

81. Da Silva Sobrinho, A.; Czeremuszkin, G.; Latreche, M.; Dennler, G.; Wertheimer, M. A study of defects in ultra-thin transparent coatings on polymers. *Surf. Coat. Technol.* **1999**, *116*, 1204–1210. [CrossRef]

82. Rossi, G.; Nulman, M. Effect of local flaws in polymeric permeation reducing barriers. *J. Appl. Phys.* **1993**, *74*, 5471–5475. [CrossRef]

83. Lee, Y.I.; Jeon, N.J.; Kim, B.J.; Shim, H.; Yang, T.Y.; Seok, S.I.; Seo, J.; Im, S.G. A low—Temperature thin—Film encapsulation for enhanced stability of a highly efficient perovskite solar cell. *Adv. Energy Mater.* **2018**, *8*, 1701928. [CrossRef]

84. Anderson, L.; Jacob, M. Effect of RF power on the optical and morphological properties of RF plasma polymerised linalyl acetate thin films. *Appl. Surf. Sci.* **2010**, *256*, 3293–3298. [CrossRef]

85. Xu, Q.F.; Wang, J.N.; Sanderson, K.D. Organic–inorganic composite nanocoatings with superhydrophobicity, good transparency, and thermal stability. *ACS Nano* **2010**, *4*, 2201–2209. [CrossRef] [PubMed]

86. Bazaka, K.; Jacob, M. Synthesis of radio frequency plasma polymerized non-synthetic Terpinen-4-ol thin films. *Mater. Lett.* **2009**, *63*, 1594–1597. [CrossRef]

87. Bazaka, K.; Jacob, M.V. Post-deposition ageing reactions of plasma derived polyterpenol thin films. *Polym. Degrad. Stab.* **2010**, *95*, 1123–1128. [CrossRef]

88. Jin, J.; Chen, S.; Zhang, J. Investigation of UV aging influences on the crystallization of ethylene-vinyl acetate copolymer via successive self-nucleation and annealing treatment. *J. Polym. Res.* **2010**, *17*, 827–836. [CrossRef]

89. Schlothauer, J.; Jungwirth, S.; Köhl, M.; Röder, B. Degradation of the encapsulant polymer in outdoor weathered photovoltaic modules: Spatially resolved inspection of EVA ageing by fluorescence and correlation to electroluminescence. *Sol. Energy Mater. Sol. Cells* **2012**, *102*, 75–85. [CrossRef]

90. Le Donne, A.; Dilda, M.; Crippa, M.; Acciarri, M.; Binetti, S. Rare earth organic complexes as down-shifters to improve Si-based solar cell efficiency. *Opt. Mater.* **2011**, *33*, 1012–1014. [CrossRef]

91. Hayes, R.A.; Lenges, G.M.; Pesek, S.C.; Roulin, J. Low Modulus Solar Cell Encapsulant Sheets with Enhanced Stability and Adhesion. United States Patent 8168885, 5 January 2012.

92. *Annual Report 2010*; Bayer AG: Leverkusen, Germany, 28 February 2011. Available online: https://www.bayer.com/en/gb-2010-en.pdfx (accessed on 29 November 2018).

93. Schut, J.H. Shining Opportunities in Solar Films. Plastics Technology. Available online: http://search.ebscohost.com/login.aspx (accessed on 29 November 2018).

94. Granstrom, J.; Swensen, J.; Moon, J.; Rowell, G.; Yuen, J.; Heeger, A. Encapsulation of organic light-emitting devices using a perfluorinated polymer. *Appl. Phys. Lett.* **2008**, *93*, 409. [CrossRef]

95. Cavalcante, L.M.; Schneider, L.F.J.; Silikas, N.; Watts, D.C. Surface integrity of solvent-challenged ormocer-matrix composite. *Dent. Mater.* **2011**, *27*, 173–179. [CrossRef]

96. Noller, K.; Vasko, K. Flexible polymer barrier films for the encapsulation of solar cells. In Proceedings of the 19th European Photovoltaic Solar Energy Conference and Exhibition 2004, Paris, France, 7–11 June 2004; pp. 2156–2159.

97. Budunoglu, H.; Yildirim, A.; Guler, M.O.; Bayindir, M. Highly transparent, flexible, and thermally stable superhydrophobic ORMOSIL aerogel thin films. *ACS Appl. Mater. Interfaces* **2011**, *3*, 539–545. [CrossRef]

98. Pagliaro, M.; Ciriminna, R.; Palmisano, G. Silica-based hybrid coatings. *J. Mater. Chem.* **2009**, *19*, 3116–3126. [CrossRef]

99. Fu, Y.; Tsai, F.-Y. Air-stable polymer organic thin-film transistors by solution-processed encapsulation. *Org. Electron.* **2011**, *12*, 179–184. [CrossRef]

100. Ong, K.S.; Raymond, G.C.R.; Ou, E.; Zheng, Z.; Ying, D.L.M. Interfacial and mechanical studies of a composite Ag–IZO–PEN barrier film for effective encapsulation of organic TFT. *Org. Electron.* **2010**, *11*, 463–466. [CrossRef]

101. Fluoropolymer. Available online: https://en.wikipedia.org/wiki/Fluoropolymer (accessed on 29 November 2018).

 *coatings* 

Review

Electroplating of Semiconductor Materials for Applications in Large Area Electronics: A Review

Ayotunde Adigun Ojo * and I. M. Dharmadasa

Electronic Materials and Sensors Group, Materials and Engineering Research Institute, Sheffield Hallam University, Sheffield S1 1WB, UK; sciimd@exchange.shu.ac.uk
* Correspondence: a8031624@my.shu.ac.uk; Tel.: +44-114-225-6910; Fax: +44-114-225-6930

Received: 26 June 2018; Accepted: 25 July 2018; Published: 27 July 2018

Abstract: The attributes of electroplating as a low-cost, simple, scalable, and manufacturable semiconductor deposition technique for the fabrication of large-area and nanotechnology-based device applications are discussed. These strengths of electrodeposition are buttressed experimentally using techniques such as X-ray diffraction, ultraviolet-visible spectroscopy, scanning electron microscopy, atomic force microscopy, energy-dispersive X-ray spectroscopy, and photoelectrochemical cell studies. Based on the results of structural, morphological, compositional, optical, and electronic properties evaluated, it is evident that electroplating possesses the capabilities of producing high-quality semiconductors usable for producing excellent devices. In this paper we will describe the progress of electroplating techniques mainly for the deposition of semiconductor thin film materials and their treatment processes, and fabrication of solar cells.

Keywords: electroplating; semiconductors; large-area electronics; characterisation; solar cells

---

1. Introduction

Electroplating has been well explored over the years especially for the purification, extraction, protection, and coating of semiconductors, metals, and metalloids in the industrial sector [1] to achieve diverse characteristic properties. The use of the electroplating technique in the deposition of semiconductor materials dates back to the 1970s [2–4], with the deposition of semiconductors from the II-VI group. The ascendance of the electrodeposition of semiconductor material led to the growth and fabrication of CdS/CdTe-based solar cell devices within a decade afterwards [5]. The fabrication of thin-film solar cells with photovoltaic conversion efficiency of ~10% was the stimulus for an intense global research into electrodeposited semiconductor compounds. The research also spanned into the electrodeposition of II-VI semiconductor materials such as ZnTe [6], ZnSe [6], ZnS [7], ZnO [8], etc., and spread into semiconductor material compounds in the binary (III-V, IV-VI), ternary (CuInSe$_2$) [9,10], and quaternary (Cu$_2$ZnSnS$_4$, CuInGaSe$_2$) groups [11]. The electroplating of elemental semiconductors and other wide bandgap nitrides has also been captured in the literature. This communication critically appraises the strengths, weaknesses, potentials, and the state-of-the-art electroplating technique in the fabrication of large-area electronics and other macro-electronic devices such as photovoltaic (PV) solar panels and display devices.

2. An Overview of Electrodeposition Technique

Electrodeposition is the process of depositing elemental or compound metals or semiconductors on a conducting substrate by passing an electric current through an ionic electrolyte in which metal or semiconductor ions are present [12]. The passage of current is required due to the inability of the chemical reaction resulting in the deposition of the solid material on the conducting substrate to proceed on its own as a result of positive free energy change ΔG of the reaction.

Electrodeposition can be categorised based on power supply source, working electrode, and electrode configuration (as shown in Figure 1), but the basic deposition mechanism and setup remains similar. The basic deposition mechanism entails the flow of electrons from the power supply to the cathode. The positively charged cations are attracted towards the cathode and negatively charged anions to the anode. The cations or anions are neutralised electrically by gaining electrons (through reduction process) or losing electrons (through oxidation process) and being deposited on the working electrode (WE), respectively [12].

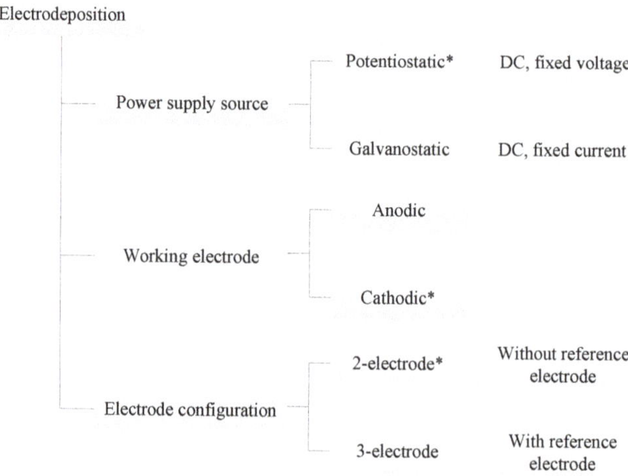

Figure 1. The main categories of electrodeposition technique.

The typical electrodeposition (ED) setup of a two-electrode (2E) configuration, as shown in Figure 2a, consists of a deposition container (beaker), deposition electrolyte, magnetic stirrer, hotplate, power supply, a working electrode, a counter electrode, and an optional reference electrode (RE) in the case of a three-electrode (3E) configuration (see Figure 2b). The use of a potentiostatic power source was due to the effect of deposition voltage on the atomic percentage composition of elements in the electrodeposited layer, which is one of the factors determining the conductivity type [13,14]. Cathodic deposition is mainly utilized due to its ability to produce stoichiometric thin-films with good adherence to the substrate as compared to anodic deposition [15]. Conversely, the galvanostatic electrodeposition is controlled and measured by maintaining constant current density through an electrolytic cell disregarding the changes in the resistance due to the deposited electroplated layer.

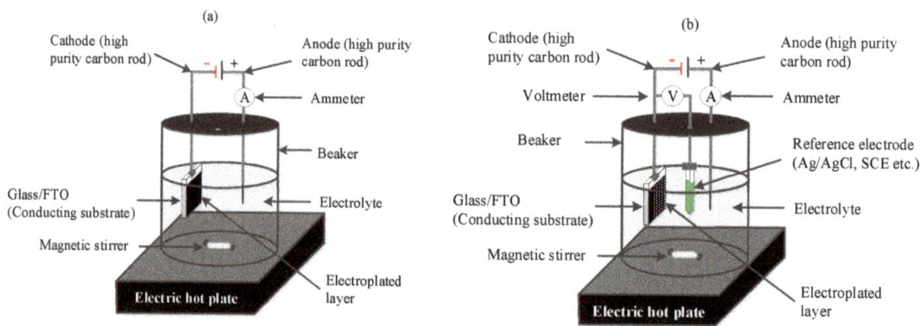

Figure 2. Typical (**a**) two-electrode and (**b**) three-electrode electrodeposition setups.

The 2E configuration, as shown in Figure 2a, was utilised due to its industrial applicability, process simplification, and also to eliminate possible Ag^+- and K^+-ion doping [16,17], which may emerge from the Ag/AgCl or saturated calomel electrode (SCE) reference electrodes (see Figure 2b). Taking the electrodeposition of n-CdS and n-CdTe layers which are respectively utilised as the main window and absorber layers in this work into perspective, both K^+ and Ag^+ from group I of the periodic table are considered as p-type dopants. Therefore, any leakage of K^+ and Ag^+ into the electrolytic bath may result in compensation leading to the growth of highly resistive material which has a detrimental effect on the efficiency of fabricated solar cells. This has been experimentally shown and reported in the literature [16].

The two-electrode electrodeposition configurations are not without challenges, with the main challenge being the fluctuation or drop in the potential measured across the cathode and the anode during deposition. This is due to the alteration in resistivity of the substrate with increasing semiconductor layer thickness and the change in the ionic concentration of the electrolyte. For the three-electrode configuration, the potential difference is measured across the working and the reference electrodes, while the measured current is between the working and the counter electrodes. In general, other factors such as the pH of the electrolyte [18], applied deposition potential [13,14], deposition temperature [19], stirring rate [20], deposition current density, duration of deposition and thickness [21], underlying substrate [22], and concentration of ions in the deposition electrolyte [18] affects the electrodeposition process and the properties of the deposited layers. Recent publications have demonstrated the similarities between electrodeposited semiconductors using three-electrode and two-electrode electroplating configurations [23,24]. The electrodeposition of both elements and compounds is governed by Faraday's laws of electrodeposition as mathematically depicted in Equation (1).

$$T = \left(\frac{1}{nF}\right)\left(\frac{itM}{\varrho A}\right) = \left(\frac{JtM}{nF\varrho}\right) \tag{1}$$

where T is the thickness (cm), J is current density (A cm^{-2}), t is the deposition time (s), M is the molecular mass (g mol^{-1}), n is the number of electrons transferred in the chemical reaction for the formation of 1 mole of substance in g cm^{-3}, F is Faraday's constant (96,485 C mol^{-1}), and ϱ is the density (g cm^{-3}). It should be noted that Faraday's law of electrolysis assumes that all electronic charges passing through the electrolyte contribute to the deposition of deposited material layer without any consideration of the resistance losses in the system and electronic charge contribution to the decomposition of solvent into its constituent ions [25].

3. Factors Influencing Electrodeposition

3.1. Solutes, Solvents, and Deposition Electrolytes

The effects of the incorporated solute and solvent utilised are of importance in electrodeposition. Taking the electrodeposition of CdS into consideration, sodium-(Na)-based precursor ($Na_2S_2O_3$) has been often utilised [26,27]. Although sodium (Na) ions are not electrodeposited at low cathodic voltages, the incorporation of Na in CdS films is achievable through adsorption, absorption or chemical reactions as a result of increased Na accumulation in the electrolytic bath. It should be noted that Na is a p-type dopant in CdS [28] resulting in increasing electrical resistivity of subsequent CdS layers due to Na accumulation. Further to this, the Na-based precursor ($Na_2S_2O_3$) is also associated with the precipitation of sulphur during the electroplating. Recent understanding has shown that the replacement of the well-established sulphur precursor with thiourea (NH_2CSNH_2) (which is more associated with chemical bath deposition (CBD) technique) results in the reduction/elimination of sulphur precipitate [29,30].

The choice of solvent to be utilised also possesses as an important factor in electroplating as demonstrated by the deposition of CdTe from ethylene glycol ($C_2H_6O_2$) electrolyte containing cadmium iodide (CdI_2) [31]. Using aqueous solution as solute, Cd-I complexes such as CdI^+, CdI_2,

CdI_3^-, and CdI_4^{2-} are formed in aqueous solution [32,33] debarring the deposition of Cd and the co-deposition of CdTe. It is noteworthy that CdTe from other Cd precursors have been explored and reported in the literature [34,35]. Therefore, the choice of solute and solvent for electroplating purposes is a factor to reckon with in addition to a number of other factors inherent in the electrodeposition process to achieve superior qualities of electroplated semiconductor materials. This understanding has been accrued for over two decades of exploration, and careful examination of grown semiconductors at Solar Energy Group within Sheffield Hallam University (SHU), in addition to the literature. A large number of semiconductors explored in SHU and documented in the literature [9,29,36–47] is summarised in Table 1.

Table 1. Summary of explored electronic materials to date at authors' research group using electroplating from aqueous solutions.

Material Electroplated	E_g (eV)	Precursors Used for Electroplating	Comments	Ref.
CuInSe$_2$	~1.00	CuSO$_4$ for Cu ions, In$_2$(SO$_4$)$_3$ for In ions and H$_2$SeO$_3$ for Se ions	Ability to grow both p- and n-type material	[9]
CdTe	1.45	CdSO$_4$ or Cd(NO$_3$)$_2$ or CdCl$_2$ for Cd ions and TeO$_2$ for Te ions	Ability to grow both p- and n-type CdTe using Cd-sulphate, nitrate, and chloride precursors	[40,41]
CuInGaSe$_2$	1.00–1.70	CuSO$_4$ for Cu ions, In$_2$(SO$_4$)$_3$ for In ions, Ga$_2$(SO$_4$)$_3$ for Ga ions and H$_2$SeO$_3$ for Se ions	Ability to grow both p- and n-type material	[42]
CdSe	1.90	CdCl$_2$ for Cd ions and SeO$_2$ for Se ions	–	[43]
InSe	1.90	InCl$_3$ for In ions and SeO$_2$ for Se ions	–	[44]
GaSe	2.00	Ga$_2$(SO$_4$)$_2$ for Ga ions and SeO$_2$ for Se ions	–	–
ZnTe	1.90–2.60	ZnSO$_4$ for Zn ions and TeO$_2$ for Te ions	Ability to grow both p- and n-type material	[45]
CdS	2.42	CdCl$_2$ for Cd ions and Na$_2$S$_2$O$_3$, NH$_4$S$_2$O$_3$ or NH$_2$CSNH$_2$ for S	Conductivity type is always n-type	[29,46,47]
CdMnTe	1.57–2.50	CdSO$_4$ for Cd ions, MnSO$_4$ for Mn ions and TeO$_2$ for Te ions	–	–
ZnSe	2.70	ZnSO$_4$ for Zn ions and SeO$_2$ for Se ions	Ability to grow both p- and n-type material	[36]
ZnO	3.30	Zn(NO$_3$)$_2$ for Zn ions	–	[37]
ZnS	3.75	ZnSO$_4$ for Zn and (NH$_4$)$_2$S$_2$O$_3$ for S ions	Ability to grow both p- and n-type material	[38]
Polyaniline (PAni)	–	C$_6$H$_5$NH$_2$ and H$_2$SO$_4$	To use as a pinhole plugging layer	[39]

3.2. Electrolytic Bath pH Value

The composition of an electrolytic bath naturally determines the pH of the bath. Basically, the acidity (pH < 7.00) of an electrolyte can be increased by the introduction of an acid. The hydrogen ions (H$^+$) from the dissociated acid reacts with water in aqueous solution to form hydronium ions (H$_3$O$^+$). On the other hand, the alkalinity of a solution increases (pH > 7.00) with the reduction in the H$_3$O$^+$ concentration. This is caused by the reaction of dissociated hydroxide ions (OH$^-$) from introduced alkaline with H$^+$ ions from water dissociation to form water (H$_2$O) rather than hydronium ions. It is well documented that elemental and compound deposition responds to this chemical dynamic mainly in wet deposition techniques such as chemical bath deposition (CBD) [48] and electrodeposition techniques [49]. With emphasis on electrodeposition, the effect of pH on the bath and the deposited layers vary from selective deposition/etching of element [50], alteration of the characteristic properties of the deposited layers [51,52], elemental/compound precipitation [53], and increase in the deposition current density [54]. Furthermore, the effect of pH on the dissociation of common solvent such as water is also well documented in the literature [55], with the notion that an increase in the acidity of an electroplating bath results in the increase in the concentration of dissociated ions in the aqueous solution [55]. Due to the increased ionic concentration, the deposition current density increases until it stabilises or continues to increase, depending on the composition of the solution.

3.3. Deposition Temperature

It is a known fact that an increase in the temperature of a matter increases the motion of the molecules inside it. As such, the electrolytic bath temperature increases solubility of the solvent, catalyses the reactions, energizes the ions, and increases the transport number, which results in an increase in the deposition current density and rate of deposition of constituent elements or compounds. Further to this, the work performed on electrodeposition depicted that an increase in the crystallinity of as-deposited semiconductor material is achievable at higher growth temperature [3,56]. For electroplated semiconductor materials from aqueous solution, there is a limitation on the growth temperature due to the boiling temperature of water at 100 °C under standard atmospheric pressure, while the electroplating from other electrolytic baths can go as high as 160 °C [31]. Deposition of materials at higher temperature provides energy required for ions/atoms to move around and deposit in a regular crystalline pattern.

3.4. Deposition Current Density

With regards to Faraday's law of electrodeposition, the deposition current density is directly related to the thickness of the deposited layer. Thus, the deposition current density is dependent on factors effecting the energizing of the inherent ions in the electrolyte such as stirring rate, bath temperature, concentration of constituent [57], and electrical conductivity of the substrate amongst other constraints. While a gradual alteration in the deposition current density is expected depending on the electrical conductivity of the electroplated layers.

With respect to semiconductor materials such as CdTe, the literature depicts the effect of current density on the morphological, compositional and the structural properties of the deposited layer [58,59]. Based on the deposition configuration, it can be inferred that the deposition current density of three-electrode configuration and two-electrode deposition configuration vary. CdTe with optimal characteristic properties deposited from a three-electrode deposition is known to lie between ~(0.3–0.6) mA cm^{-2} [57,60]. While two-electrode electrodeposition has been documented to produce CdTe layers with an optimal characteristic property of ~(0.15–0.18) mA cm^{-2} [61]. Under potentiostatic mode, Basol (1988) clarified that the deposition current density for CdTe electroplating depends on the tellurium concentration in the electrolyte [57]. The incorporation of excessive Te can alter the composition of the deposited CdTe, conduction type to p-CdTe due to Te-richness [62], and reduce adhesion on the underlying substrate. Reduction in the adhesion of CdTe may also occur due to the deficiency of Te concentration in the electrolytic bath. In either condition (excess or deficiency of Te concentration in the electrolyte), the crystallinity, morphology, and adhesion of the ensued CdTe layer suffer. In addition, current density during co-deposition of competing ions such as Ga^{3+} and Fe^{2+} in the electroplating of Fe–Ga alloys is a determinant of the Fe:Ga elemental composition ratio of the resultant alloy [63].

3.5. Duration of Deposition and Thickness

Electroplating of materials with main emphasis on semiconductor commences by the nucleation of the most electropositive element on the points on the conducting substrates with the highest electric field. Therefore, it can be categorically stated that the nucleation and nucleation modes of semiconductor material is conductive substrate dependent [1,22,64]. Consequential to the surface roughness of the underlying working electrode such as glass/fluorine-doped tin oxide (g/FTO), the highest electric field is experienced at the peaks of the rough surfaces. The nucleation of the electroplated material spreads out through to the lowest valley from the initiation rough surface peaks resulting into columnar nature of the deposited layers [14]. The potency of this mechanism is highly influential at the initial stages of deposition due to unevenness of the deposited layer thickness characterised by pin-holes, voids, gaps, and high dislocation density within the semiconductor material [21]. This characteristic property is detrimental when a thin semiconductor layer with a thickness of <100 nm is required [21].

4. Strengths and Weaknesses of Electrodeposition

4.1. Strengths of Electrodeposition

4.1.1. Electrolytic Bath Life Longevity and Self -Purification

At the start of an electrolytic bath, electro-purification of the bath is highly essential to reduce and eliminate the impurity level, which is mostly incorporated in the precursors amongst other impurity sources. It should be noted that even with a high purity precursor with 99.999% purity, it can carry an impurity level of 10 ppm. Purification is essential due to the effect of impurities even in ppm levels [65] on the characteristic properties of electroplated semiconductor materials. It should be noted that electro-purification of a bath must be performed using similar deposition parameters (such as bath temperature, pH, stirring rate, etc.) to the semiconductor deposition. The electro-purification potential utilised should be lower than the deposition potential range of the required elements established using cyclic voltammetry. Based on this characteristic property of electroplating technique, the more layers deposited, the purer the electrolyte and the electroplated semiconductor gets due to the gradual reduction of background impurities and improved material property. This property not only increases the purity of the electrolyte and the deposited semiconductor, but also increases the longevity of the bath as compared to the batch process of chemical bath deposition (CBD) technique.

To further mitigate other sources of impurities, a fraction of researchers choose two-electrode over the three-electrode configuration to avoid possible impurities from the reference electrode. While the usage of Teflon-ware (polypropylene beaker) is necessitated to house the electrolyte due to possible leaching of elemental sodium and other dopants from glass-wares [66] into acidic electrolytes.

4.1.2. Ease of Doping Intrinsic and Extrinsic

With the effective purification of the electrolyte, intrinsic doping has been demonstrated in the literature for binary [38,67], ternary [68], and quaternary [69,70] semiconductor materials by changing the deposition voltage. Taking an example of a I–III–VI$_2$ semiconductor materials such as CuInGaSe$_2$, the stoichiometric semiconductor layer consists of 25% of the group I element, 25% of the group III elements, and 50% of group VI element. Due to the positive reduction potential of Cu (E_0 = 0.52 V), at low deposition voltages, high elemental composition of Cu (group I) is incorporated in the semiconductor resulting into *p*-type conduction type. But an increase in the cathodic voltage increases the elemental composition resulting in an *n*-type semiconductor material, as in the case of CuInSe2 (see Figure 3a). While at intermediate voltages, the material exhibits insulating or intrinsic properties. This electrical characteristic property, as demonstrated in the literature [68–70], signifies the ability of growth of *p*-, *i*-, and *n*-type materials from the same bath by cathodic voltage variation (see Figure 3). The incorporation of Ga in CuInGaSe$_2$ [42] increases the bandgap and also makes the material *p*-type. This must be due to the formation of acceptor-like defects in the material (see Figure 3b).

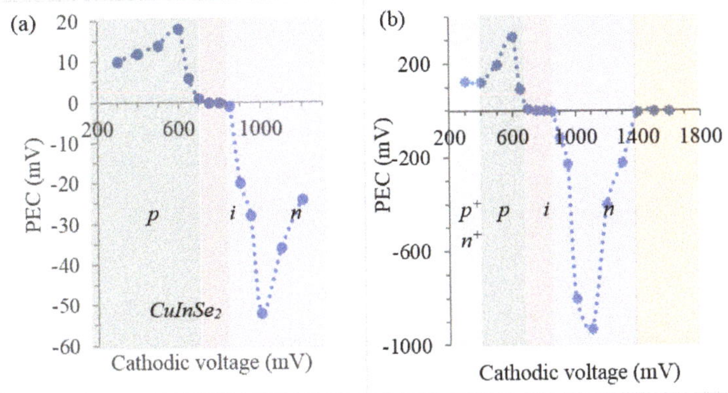

Figure 3. Photoelectrochemical (PEC) cell measurement signal for (a) CuInSe$_2$ and (b) CuInGaSe$_2$ with increasing cathodic voltages. Note the ability to grow p$^+$, p, i, n, and n$^+$ materials from the same electrolyte, simply by varying the deposition voltage. Reproduced from [68,70] with permission; Copyright 2004 Sheffield Hallam University.

The effect of cathodic voltage on the elemental composition of binary semiconductors has also been demonstrated and documented in the literature [14,62]. The effect of alteration in the growth voltage on the elemental composition of electroplated materials even for as low as 1 mV step has been documented [29] (see Figure 4).

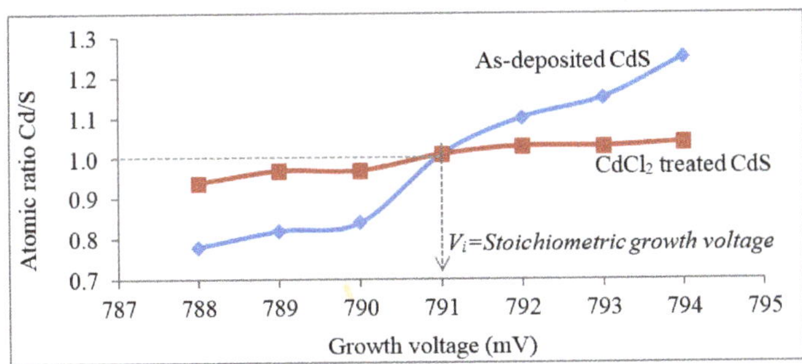

Figure 4. Atomic composition ratios of Cd to S in as-deposited and CdCl$_2$ treated CdS thin films at different deposition cathodic voltages.

The ease of intrinsic doping and the effect of extrinsic doping of electroplated semiconductor materials have been well established in the literature [65,71]. Due to the simplicity of ED, doping at the ppm level is made possible [65,71,72].

4.1.3. Bandgap Engineering Capability

The control or alteration of the bandgap of materials (with emphasis on semiconductor) is easily achievable in electrodeposition techniques. Typically, this can be achieved by controlling the atomic composition of the elemental component of the semiconductor material. Intrinsically, electrodeposition has the ability to change the composition of growing material by a simple alteration of the cathodic voltage [38,68–70]. An ensuing alteration in the bandgap of grown semiconductor material due to change in the growth cathodic voltage has been documented in the literature [42]. It is well known that an increase in the atomic concentration of Ga in CuInGaSe$_2$ by increasing the cathodic voltage increases the bandgap of CuInGaSe$_2$. While a reduction in the cathodic voltage of the CuInGaSe$_2$ results in the reduction of bandgap due to the richness of Cu [42] (see Figure 5). This ability provides the ease of bandgap engineering of semiconductor material such as CuInGaSe$_2$ between ~1.00 and 2.20 eV. Extrinsically, this observation has also been documented for electroplated binary semiconductor materials such as CdTe doped with Ga [65] amongst others. The bandgap of the resulting doped semiconductor directly affected the incorporated dopant even at the parts per million level [65,71].

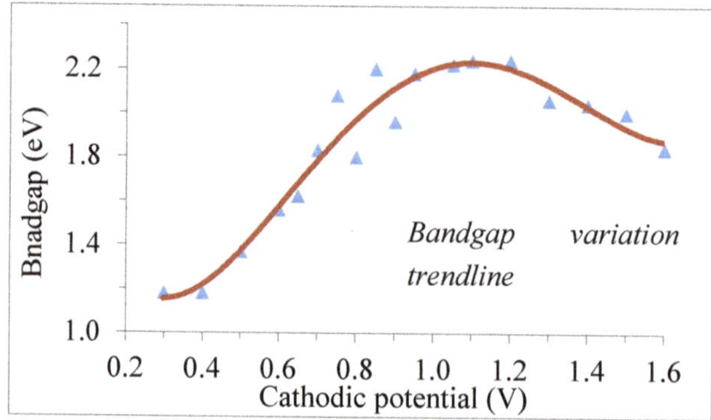

Figure 5. Energy bandgap variation for CuInGaSe$_2$ as a function of the cathodic voltage.

4.1.4. Low Cost and Simplicity

There are over 14 different and well-established techniques to grow thin-film semiconductor materials [73] which can be broadly categorised under physical or chemical deposition. Physical deposition refers to the technologies in which material is released from a source and deposited on a substrate using thermodynamic, electromechanical or mechanical processes [1,74]. Chemical deposition techniques are accomplished by the utilisation of precursors either in their liquid or gaseous state to produce a chemical reaction on the surface of a substrate, leaving behind chemically deposited thin-film coatings on the substrate. Electrodeposition falls under chemical deposition techniques which can be carried out in an uncontrolled environment and without a vacuum system. The setup for electroplating which is mainly constituted of a computerized potentiostat and hotplate/magnetic stirrer with a cost implication of less than £5000 as compared to other techniques such as the well-established metallorganic chemical vapour deposition (MOCVD) or close space sublimation (CSS) system with a high initial cost implication of about £1 million for a laboratory setup. In addition, these systems have limitations as concerning the materials that can be grown. Furthermore, the relatively low heat energy required during growth and post-growth treatment makes electroplating a more energy-economic deposition technique as compared to a large number of other techniques. More importantly, grown semiconductor layers using cost-effective electroplating techniques are comparable to semiconductor layers grown using highly expensive techniques [29,75], and they all require post-deposition treatments [76,77].

4.1.5. Scalability and Manufacturability

The scalability and manufacturability of electroplating has been demonstrated on an industrial scale by British Petroleum (BP) Solar in the 1980s and 1990s [78,79]. BP Solar manufactured CdTe-based solar cells with a solar panel area ~1 m^2 with a conversion efficiency of ~10% [78,79]. As compared to the laboratory scale setup as shown in Figure 2, scaling up requires a larger tank to contain the electrolyte and multi-plate cathode attached to multiple conducting substrates. The use of larger tanks and multi-plate cathode increases the throughput of deposited layers and an added advantage of electroplating on intricate shapes and designs.

4.2. Weaknesses of Electrodeposition

One of the main disadvantages of electrodeposition includes comparatively low rates of deposition and the need for a conducting substrate as the working electrode in the electroplating setup. Due to this requirement, using conventional characterisation techniques such as the Hall effect to determine the electrical properties of the deposited layers on FTO, for example, will not be possible due to the underlying conducting layer.

4.2.1. Instability of Current Density during Deposition

The control of the electrodeposition process due to the alteration of current density with increasing deposition layer thickness is a challenge (under potentiostatic condition). The electroplating of materials with electrical conductivity levels lower than the primary substrate results in the reduction of current density with a direct relationship with the thickness of the deposited material [80]. This observation is common for both 2E and 3E electroplating configurations, but the applied voltage can vary slightly in 2E configuration.

4.2.2. Control and Regulation of Ions within the Electrolytic Bath

The chemical concentration of ionic species in an electrolytic bath is defined during bath creation, but the control, regulation and measurement of ionic concentration within the electrolytic bath during electroplating remains a challenge. This is as a result of the depletion in the ionic concentration and the inability to gauge/measure the concentration change during and after initial deposition, therefore replenishing the bath with the appropriate chemical concentration becomes difficult and thereby reducing reproducibility tendencies of electroplated layers.

4.2.3. Formation of Solution-Based Complexes

There is a possibility for the formation of complexes within the electrolyte which might be debarring the deposition of an element and/or the co-deposition of a compound [32,33]. This is the case for the deposition of CdTe from aqueous solution containing CdI_2 as the Cd-precursor. The literature shows that due to the formation of Cd-I complexes in aqueous solution, only p-CdTe layers due to Te-richness is possible [32,33]. Unnecessary precipitation removes chemicals from the electrolyte, changing the elemental concentration in the bath.

4.2.4. Extrinsic Doping of Electrolytic Bath by the Electrodes

Control of purity throughout the electrolytic bath lifespan if an electrolytic bath is essential. It is well known that the purity of an electrolytic bath increases with the deposition aging of the bath but there is a tendency of an influx of impurity which might be due to etching, corrosion or dissolution of the electrode. For electrodeposition setup using carbon electrode, there has been observation of increased carbon concentration in deposited semiconductor layers. The incorporation of carbon into the electrolytic bath is due to the deterioration of the anode utilised in the electrolytic cell setup.

4.2.5. Non-Uniformity of Electrodeposited Semiconductor Layers

Due to the unevenness of the underlying conducting substrate such as transparent conducting oxide (TCO), the highest electric field is experienced at the peaks of the rough conducting substrate surfaces. Nucleation starts at the peaks and spreads out through to the lowest valley resulting into layers with columnar nature [14].

4.2.6. Post-Growth Treatment

It is well documented that electrodeposited layers with an emphasis on semiconductors often require post-deposition treatment to further improve their structural, morphological, compositional, and optical properties [81–83]. It is, although, arguable that other semiconductor deposition techniques such as [77], MOCVD [84], and CSS [85] also require such treatments, as it is evident in the performances of the applications utilised for References [81–83].

Asides semiconductors, other issues such as the influence of different process conditions on mechanical properties of electroplated materials and the control of the integrity of the electroplated layers as it relates to further processing have also been the raised [86,87], but not discussed in this communication.

5. All-Electroplated Photovoltaic Devices

The comparability of the structural, morphological, compositional, optical, and other material properties has been well documented in the literature [29,42,68,88]. Electrodeposited cadmium telluride (CdTe) and copper indium gallium selenide (CIGS) are amongst the commonly used absorber layers in all-electrodeposited photovoltaic applications [57,89]. The versatility of the technique in the growth of all-electrodeposited configurations has been well documented [13,90–92]. The band diagrams of possible n-p and n-n$^+$ large Schottky barrier junctions are shown in Figure 6 fabricated using CdS/CdTe configuration. The full characterisation process of both the CdS and CdTe are documented in the literature [40,46]. The electronic properties of the fabricated photovoltaic cells obtained using current–voltage (*I–V*) and capacitance–voltage (*C–V*) techniques are summarised in Table 2.

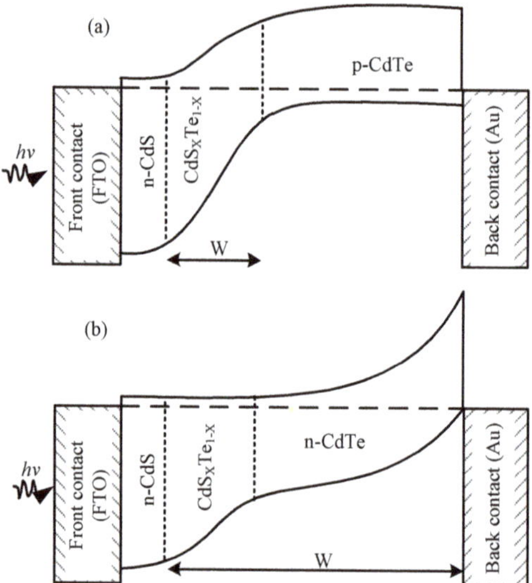

Figure 6. Energy band diagrams representing (**a**) glass/FTO/n-CdS/p-CdTe/Au and (**b**) glass/FTO/ n-CdS/n-CdTe/Au device configurations.

Table 2. Summary of device parameters obtained from *I–V* (both under illuminated and dark conditions) and *C–V* (dark condition) for simple CdS/CdTe-based solar cells grown at different growth voltages in the vicinity of V_i = 1370 mV.

Parameters	Values				
CdTe Growth Voltage (mV)	1340	1360	1370	1380	1400
I–V under dark condition					
R_{sh} (Ω)	1016	>10^5	>10^5	>10^5	>10^5
R_s (kΩ)	0.21	0.80	0.50	1.43	1.50
log (RF)	0.4	3.5	3.9	3.3	3.0
I_o (A)	2.5×10^{-5}	3.9×10^{-9}	1.0×10^{-9}	3.2×10^{-9}	5.0×10^{-9}
n	>2.00	1.95	1.86	1.58	1.86
Φ_b (eV)	>0.52	>0.76	>0.81	>0.77	>0.77
I–V under AM1.5 illumination condition					
I_{sc} (mA)	0.53	0.62	0.65	0.82	0.57
J_{sc} (mA cm^{-2})	16.88	19.75	20.70	26.11	18.15
V_{oc} (V)	0.23	0.49	0.72	0.60	0.57
Fill factor	0.31	0.46	0.50	0.45	0.48
Efficiency (%)	1.20	4.45	7.50	7.05	4.97
C–V under dark condition					
$\sigma \times 10^{-4}$ (Ω cm^{-1})	1.41		2.85		6.03
N_A or N_D (cm^{-3})	7.74×10^{16}		3.10×10^{14}		9.10×10^{14}
μ (cm^2 V^{-1} s^{-1})	0.01	–	5.74	–	4.14
C_o (pF)	1630		330		370
W (nm)	187.6		926.7		826.5

It is well known that intrinsic CdS is n-type and remains *n*-type due to the inherent defect as a result of the presence of S vacancies and Cd interstitials in the crystal lattice of the deposited CdS layers [93]. The devices are fabricated by incorporating CdTe deposited at the vicinity of the transition voltage (V_i) from p-type to n-type CdTe or vice versa. Electroplated CdTe can either be p-type (when Te rich) or n-type (when Cd rich) material under as-deposited condition. Retention or transition of electrical conduction type is possible after cadmium chloride treatment. It is noteworthy that the conversion of the electrical conduction type after post-growth treatment may be attributed to the doping effect as a result of the heat treatment temperature, duration of treatment, initial atomic composition of Cd and Te, the concentration of CdCl$_2$ utilised in treatment, defect structure present in the starting CdTe layer, and the material's initial conductivity type as documented in the literature [40,41,94]. Therefore depending on the final electrical conduction type, the possible device configurations are possible, and the analysis of device results must be performed with extreme care.

Table 2 summarises the results of CdS/CdTe solar cells made with CdTe layers grown in the vicinity of V_i = 1370 mV. Below the V_i, the CdTe layers are p-type and therefore the devices made are p-n junctions (see Figure 6a). Above the V_i, the CdTe layers are n-type and hence the device structures are n-n$^+$ Schottky barrier (see Figure 6b). As shown in Table 2, the devices fabricated with n-CdTe performs better than those made with p-CdTe layers.

CdTe-based devices assuming p-CdTe in CdS/CdTe devices has been documented in the literature [27,95]. But based on recent observations, the incorporation of Cd-rich CdTe absorber layers produce high efficiencies. These effects have been independently observed and reported [96,97] and mainly attributed to the reduced defects in Cd-rich CdTe (the layers are deposited using physical deposition processes).

Using mainly n-CdTe absorber layers, few devices incorporating all-electrodeposited from the SHU group have been documented in the literature and summarised in Table 3.

Table 3. Summary of device parameters obtained from *I–V* (both under illuminated and dark conditions) and *C–V* (dark condition) for glass/FTO/n-CdS/n-CdTe/Au and glass/FTO/n-ZnS/n-CdS/n-CdTe/Au solar cells.

Properties	Configuration			
	g/FTO/n-CdS/ n-CdTe/Au (Two-Layer Device)	g/FTO/n-CdS/ n-CdTe/p-CdTe/ Au (Three-Layer Device)	g/FTO/n-CdS/ n-CdTe/p-CdTe/ Cu-Au (Three-Layer Device)	Glass/FTO/n-ZnS/ n-CdS/n-CdTe/Au (Three-Layer Device)
I–V under dark condition				
R_{sh} (Ω)	$>10^5$	$>7.2 \times 10^5$	10^6	$>10^5$
R_s (kΩ)	0.50	0.50	0.92	0.47
log (RF)	3.9	4.1	3.5	4.8
I_o (A)	1.0×10^{-9}	1.0×10^{-9}	3.16×10^{-9}	1.0×10^{-9}
n	1.86	1.86	1.68	1.60
Φ_b (eV)	>0.81	>0.80	>0.80	>0.82
I–V under 1.5 AM illumination condition				
I_{sc} (mA)	0.65	1.06	1.85	1.07
J_{sc} (mA cm^{-2})	20.70	33.80	58.9	34.08
V_{oc} (V)	0.72	0.73	0.64	0.73
Fill factor	0.50	0.62	0.50	0.57
Efficiency (%)	7.50	15.3	18.5	14.18
C–V under dark condition				
$\sigma \times 10^{-4}$ (Ω cm)$^{-1}$	2.85	–	–	8.82
N_D-N_A (cm^{-3})	3.10×10^{14}	6.67×10^{14}	1.82×10^{14}	7.79×10^{14}
μ (cm^2 V^{-1} s^{-1})	5.74	–	–	7.07
C_o (pF)	330	395	160	280
W (nm)	926.7	–	–	1092.2

6. Conclusions

This work describes electroplating as a robust material deposition technique with wide applications ranging from surface protection to large-area electronics and nano-technology while focusing on semiconductor deposition. The manuscript also reviews the pros and cons of electroplating techniques. The effect of growth parameters such as temperature, pH, stirring rate, precursor, solvent and cathodic voltage, and post-growth heat treatments of the deposited were iterated. The capability of electroplated material to be comparable and possibly superior to semiconductor materials grown using other cash intensive techniques are also highlighted with experimental evidence. Electroplated materials can be applicable in large-area devices such as photovoltaic solar panels and large-area display panels in which intricate shapes are required. Bandgap grading, alteration of elemental composition, and different conductivity type are also possible intrinsically with a change in the cathodic voltage. Other advantages such as columnar growth of nanorods which are tightly packed and normal to the substrate could trigger many new applications in the nanotechnology area.

Funding: This research received no external funding

Acknowledgments: The authors acknowledge the support of Sheffield Hallam University and Ekiti State University.

References

1.	Lincot, D. Electrodeposition of semiconductors. *Thin Solid Films* **2005**, *487*, 40–48. [CrossRef]
2.	Kroöger, F.A. Cathodic deposition and characterization of metallic or semiconducting binary alloys or compounds. *J. Electrochem. Soc.* **1978**, *125*, 2028–2034. [CrossRef]

3. Panicker, M.P.R. Cathodic deposition of CdTe from aqueous electrolytes. *J. Electrochem. Soc.* **1978**, *125*, 566–572. [CrossRef]

4. Danaher, W.J.; Lyons, L.E.; Morris, G.C. Some properties of thin films of chemically deposited cadmium sulphide. *Sol. Energy Mater.* **1985**, *12*, 137–148. [CrossRef]

5. Basol, B.M. High-efficiency electroplated heterojunction solar cell. *J. Appl. Phys.* **1984**, *55*, 601–603. [CrossRef]

6. Neumann-Spallart, M.; Königstein, C. Electrodeposition of zinc telluride. *Thin Solid Films* **1995**, *265*, 33–39. [CrossRef]

7. Sanders, B.W.; Kitai, A.H. The electrodeposition of thin film zinc sulphide from thiosulphate solution. *J. Cryst. Growth* **1990**, *100*, 405–410. [CrossRef]

8. Izaki, M. Electrolyte optimization for cathodic growth of zinc oxide films. *J. Electrochem. Soc.* **1996**, *143*, L53–L55. [CrossRef]

9. Dharmadasa, I.M.; Burton, R.P.; Simmonds, M. Electrodeposition of CuInSe$_2$ layers using a two-electrode system for applications in multi-layer graded bandgap solar cells. *Sol. Energy Mater. Sol. Cells* **2006**, *90*, 2191–2200. [CrossRef]

10. Al-Bassam, A.A.I. Electrodeposition of CuInSe$_2$ thin films and their characteristics. *Phys. B Condens. Matter* **1999**, *266*, 192–197. [CrossRef]

11. Dharmadasa, I.M. Latest developments in CdTe, CuInGaSe$_2$ and GaAs/AlGaAs thin film PV solar cells. *Curr. Appl. Phys.* **2009**, *9*, e2–e6. [CrossRef]

12. McHardy, J.; Ludwig, F. *Electrochemistry of Semiconductors and Electronics: Processes and Devices*; William Andrew: Norwich, NY, USA, 1992; ISBN 9780815513018.

13. Ojo, A.A.; Dharmadasa, I.M. 15.3% efficient graded bandgap solar cells fabricated using electroplated CdS and CdTe thin films. *Sol. Energy* **2016**, *136*, 10–14. [CrossRef]

14. Dharmadasa, I.M.; Bingham, P.; Echendu, O.K.; Salim, H.I.; Druffel, T.; Dharmadasa, R.; Sumanasekera, G.; Dharmasena, R.; Dergacheva, M.B.; Mit, K.; et al. Fabrication of CdS/CdTe-based thin film solar cells using an electrochemical technique. *Coatings* **2014**, *4*, 380–415. [CrossRef]

15. Pandey, J. Solar cell harvesting: Green renewable technology of future introduction. *Int. J. Adv. Res. Eng. Appl. Sci.* **2015**, *4*, 93–103.

16. Dennison, S. Dopant and impurity effects in electrodeposited CdS/CdTe thin films for photovoltaic applications. *J. Mater. Chem.* **1994**, *4*, 41–46. [CrossRef]

17. Zanio, K. *Semiconductors and Semimetals*; Academic Press: Cambridge, MA, USA, 1978.

18. Meulenkamp, E.A.; Peter, L.M. Mechanistic aspects of the electrodeposition of stoichiometric CdTe on semiconductor substrates. *J. Chem. Soc. Trans.* **1996**, *92*, 4077–4082. [CrossRef]

19. Shenouda, A.Y.; El Sayed, M. Electrodeposition, characterization and photo electrochemical properties of CdSe and CdTe. *Ain Shams Eng. J.* **2015**, *6*, 341–346. [CrossRef]

20. Atapattu, H.Y.R.; De Silva, D.S.M.; Pathiratne, K.A.S.; Dharmadasa, I.M. Effect of stirring rate of electrolyte on properties of electrodeposited CdS layers. *J. Mater. Sci. Mater. Electron.* **2016**, *27*, 5415–5421. [CrossRef]

21. Ojo, A.A.; Salim, H.I.; Olusola, O.I.; Madugu, M.L.; Dharmadasa, I.M. Effect of thickness: A case study of electrodeposited CdS in CdS/CdTe based photovoltaic devices. *J. Mater. Sci. Mater. Electron.* **2017**, *28*, 3254–3263. [CrossRef]

22. De Alwis, A.C.S.; Atapattu, H.Y.R.; De Silva, D.S.M. Influence of the type of conducting glass substrate on the properties of electrodeposited CdS and CdTe thin films. *J. Mater. Sci. Mater. Electron.* **2018**. [CrossRef]

23. Dharmadasa, I.; Madugu, M.; Olusola, O.; Echendu, O.; Fauzi, F.; Diso, D.; Weerasinghe, A.; Druffel, T.; Dharmadasa, R.; Lavery, B.; et al. Electroplating of CdTe thin films from cadmium sulphate precursor and comparison of layers grown by 3-electrode and 2-electrode systems. *Coatings* **2017**, *7*, 17. [CrossRef]

24. Echendu, O.K.; Okeoma, K.B.; Oriaku, C.I.; Dharmadasa, I.M. Electrochemical deposition of CdTe semiconductor thin films for solar cell application using two-electrode and three-electrode configurations: A comparative study. *Adv. Mater. Sci. Eng.* **2016**, *2016*, 3581725. [CrossRef]

25. Sun, J.; Zhong, D.K.; Gamelin, D.R. Composite photoanodes for photoelectrochemical solar water splitting. *Energy Environ. Sci.* **2010**, *3*, 1252–1261. [CrossRef]

26. Yamaguchi, K.; Yoshida, T.; Sugiura, T.; Minoura, H. A novel approach for CdS thin-film deposition: electrochemically induced atom-by-atom growth of CdS thin films from acidic chemical bath. *J. Phys. Chem. B* **1998**, *102*, 9677–9686. [CrossRef]

27. Das, S.K.; Morris, G.C. Preparation and characterisation of electrodeposited *n*-CdS/*p*-CdTe thin film solar cells. *Sol. Energy Mater. Sol. Cells* **1993**, *28*, 305–316. [CrossRef]

28. Daniel, L.; Michel, F.; Hubert, C. Chemical Deposition of Chalcogenide Thin Films from Solution. In *Advances in Electrochemical Science and Engineering*; Alkire, R.C., Kolb, D.M., Eds.; Wiley-Blackwell: Hoboken, NJ, USA, 2008; pp. 165–235. ISBN 9783527616800.

29. Ojo, A.A.; Dharmadasa, I.M. Investigation of electronic quality of electrodeposited cadmium sulphide layers from thiourea precursor for use in large area electronics. *Mater. Chem. Phys.* **2016**, *180*, 14–28. [CrossRef]

30. Salim, H.I.; Olusola, O.I.; Ojo, A.A.; Urasov, K.A.; Dergacheva, M.B.; Dharmadasa, I.M. Electrodeposition and characterisation of CdS thin films using thiourea precursor for application in solar cells. *J. Mater. Sci. Mater. Electron.* **2016**, *27*, 6786–6799. [CrossRef]

31. Chaure, N.B.; Samantilleke, A.P.; Dharmadasa, I.M. The effects of inclusion of iodine in CdTe thin films on material properties and solar cell performance. *Sol. Energy Mater. Sol. Cells* **2003**, *77*, 303–317. [CrossRef]

32. Paterson, R.; Anderson, J.; Anderson, S.S. Lutfullah transport in aqueous solutions of group IIB metal salts at 298.15 K. Part 2—Interpretation and prediction of transport in dilute solutions of cadmium iodide: An irreversible thermodynamic analysis. *J. Chem. Soc. Faraday Trans. 1 Phys. Chem. Condens. Phases* **1977**, *73*, 1773–1778. [CrossRef]

33. Paterson, R.; Devine, C. Transport in aqueous solutions of group IIB metal salts. Part 7—Measurement and prediction of isotopic diffusion coefficients for iodide in solutions of cadmium iodide. *J. Chem. Soc. Faraday Trans. 1 Phys. Chem. Condens. Phases* **1980**, *76*, 1052–1061. [CrossRef]

34. Dharmadasa, I.M.; Echendu, O.K.; Fauzi, F.; Abdul-Manaf, N.A.; Olusola, O.I.; Salim, H.I.; Madugu, M.L.; Ojo, A.A. Improvement of composition of CdTe thin films during heat treatment in the presence of $CdCl_2$. *J. Mater. Sci. Mater. Electron.* **2017**, *28*, 2343–2352. [CrossRef]

35. Dharmadasa, I.M.; Echendu, O.K.; Fauzi, F.; Abdul-Manaf, N.A.; Salim, H.I.; Druffel, T.; Dharmadasa, R.; Lavery, B. Effects of $CdCl_2$ treatment on deep levels in CdTe and their implications on thin film solar cells: A comprehensive photoluminescence study. *J. Mater. Sci. Mater. Electron.* **2015**, *26*, 4571–4583. [CrossRef]

36. Samantilleke, A.P.; Boyle, M.H.; Young, J.; Dharmadasa, I.M. Electrodeposition of *n*-type and *p*-type ZnSe thin films for applications in large area optoelectronic devices. *J. Mater. Sci. Mater. Electron.* **1998**, *9*, 231–235. [CrossRef]

37. Wellings, J.S.; Chaure, N.B.; Heavens, S.N.; Dharmadasa, I.M. Growth and characterisation of electrodeposited ZnO thin films. *Thin Solid Films* **2008**, *516*, 3893–3898. [CrossRef]

38. Madugu, M.L.; Olusola, O.I.-O.; Echendu, O.K.; Kadem, B.; Dharmadasa, I.M. Intrinsic doping in electrodeposited ZnS thin films for application in large-area optoelectronic devices. *J. Electron. Mater.* **2016**, *45*, 2710–2717. [CrossRef]

39. Abdul-Manaf, N.A.; Echendu, O.K.; Fauzi, F.; Bowen, L.; Dharmadasa, I.M. Development of polyaniline using electrochemical technique for plugging pinholes in cadmium sulfide/cadmium telluride solar cells. *J. Electron. Mater.* **2014**, *43*, 4003–4010. [CrossRef]

40. Salim, H.I.; Patel, V.; Abbas, A.; Walls, J.M.; Dharmadasa, I.M. Electrodeposition of CdTe thin films using nitrate precursor for applications in solar cells. *J. Mater. Sci. Mater. Electron.* **2015**, *26*, 3119–3128. [CrossRef]

41. Abdul-Manaf, N.A.; Salim, H.I.; Madugu, M.L.; Olusola, O.I.; Dharmadasa, I.M. Electro-plating and characterisation of CdTe thin films using $CdCl_2$ as the cadmium source. *Energies* **2015**, *8*, 10883–10903. [CrossRef]

42. Dharmadasa, I.M.; Chaure, N.B.; Tolan, G.J.; Samantilleke, A.P. Development of p^+, p, i, n, and n^+-Type $CuInGaSe_2$ layers for applications in graded bandgap multilayer thin-film solar cells. *J. Electrochem. Soc.* **2007**, *154*, H466–H471. [CrossRef]

43. Olusola, O.I.; Echendu, O.K.; Dharmadasa, I.M. Development of CdSe thin films for application in electronic devices. *J. Mater. Sci. Mater. Electron.* **2015**, *26*, 1066–1076. [CrossRef]

44. Madugu, M.L.; Bowen, L.; Echendu, O.K.; Dharmadasa, I.M. Preparation of indium selenide thin film by electrochemical technique. *J. Mater. Sci. Mater. Electron.* **2014**, *25*, 3977–3983. [CrossRef]

45. Olusola, O.I.; Madugu, M.L.; Abdul-Manaf, N.A.; Dharmadasa, I.M. Growth and characterisation of n- and p-type ZnTe thin films for applications in electronic devices. *Curr. Appl. Phys.* **2016**, *16*, 120–130. [CrossRef]

46. Abdul-Manaf, N.A.; Weerasinghe, A.R.; Echendu, O.K.; Dharmadasa, I.M. Electro-plating and characterisation of cadmium sulphide thin films using ammonium thiosulphate as the sulphur source. *J. Mater. Sci. Mater. Electron.* **2015**, *26*, 2418–2429. [CrossRef]

47. Diso, D.G.; Muftah, G.E.A.; Patel, V.; Dharmadasa, I.M. Growth of CdS layers to develop all-electrodeposited CdS/CdTe thin-film solar cells. *J. Electrochem. Soc.* **2010**, *157*, H647–H651. [CrossRef]

48. Ben Nasr, T.; Kamoun, N.; Kanzari, M.; Bennaceur, R. Effect of pH on the properties of ZnS thin films grown by chemical bath deposition. *Thin Solid Films* **2006**, *500*, 4–8. [CrossRef]

49. Wu, W.; Eliaz, N.; Gileadi, E. The effects of pH and temperature on electrodeposition of Re-Ir-Ni coatings from aqueous solutions. *J. Electrochem. Soc.* **2014**, *162*, D20–D26. [CrossRef]

50. Kum, M.C.; Yoo, B.Y.; Rheem, Y.W.; Bozhilov, K.N.; Chen, W.; Mulchandani, A.; Myung, N.V. Synthesis and characterization of cadmium telluride nanowire. *Nanotechnology* **2008**, *19*, 325711. [CrossRef] [PubMed]

51. Londhe, P.U.; Chaure, N.B. Effect of pH on electrodeposited ZnO thin films. *AIP Conf. Proc.* **2012**, *1447*, 671–672. [CrossRef]

52. Kafi, F.S.B.; Jayathileka, K.M.D.C.; Wijesundera, R.P.; Siripala, W. Effect of bath pH on interfacial properties of electrodeposited n-Cu$_2$O films. *Phys. Status Solidi b* **2018**, *255*, 1700541. [CrossRef]

53. Rami, M.; Benamar, E.; Fahoume, M.; Ennaoui, A. Growth analysis of electrodeposited CdS on ITO coated glass using atomic force microscopy. *Phys. Status Solidi a* **1999**, *172*, 137–147. [CrossRef]

54. Haerifar, M.; Zandrahimi, M. Effect of current density and electrolyte pH on microstructure of Mn–Cu electroplated coatings. *Appl. Surf. Sci.* **2013**, *284*, 126–132. [CrossRef]

55. Hillel, D. Water properties. In *Encyclopedia of Soils in the Environment*; Academic Press: Cambridge, MA, USA, 2005; pp. 290–300.

56. Huang, C.; Chuang, C.; Hsu, F.; Li, K. The properties of ZnO deposits electroplated on an ITO glass substrate with direct- and pulse currents. *Int. J. Mater. Mech. Manuf.* **2017**, *5*, 1–5. [CrossRef]

57. Basol, B.M. Electrodeposited CdTe and HgCdTe solar cells. *Sol. Cells* **1988**, *23*, 69–88. [CrossRef]

58. Rastogi, A.C.; Balakrishnan, K.S. Growth, structure and composition of electrodeposited CdTe thin films for solar cells. *Sol. Energy Mater. Sol. Cells* **1995**, *36*, 121–146. [CrossRef]

59. Razmjoo, O.; Bahrololoom, M.E.; Najafisayar, P. The effect of current density on the composition, structure, morphology and optical properties of galvanostatically electrodeposited nanostructured cadmium telluride films. *Ceram. Int.* **2017**, *43*, 121–127. [CrossRef]

60. Donghwan, K.; Pozder, S.; Zhu, Y.; Trefny, J.U. Polycrystalline thin film CdTe solar cells fabricated by electrodeposition. In Proceedings of the 1994 IEEE 1st World Conference on Photovoltaic Energy Conversion—WCPEC (A Joint Conference of PVSC, PVSEC and PSEC), Waikoloa, HI, USA, 5–9 December 2014; pp. 334–337.

61. Ojo, A.A. Engineering of Electroplated Materials for Multilayer Next Generation Graded Bandgap Solar Cells. Ph.D. Thesis, Sheffield Hallam University, Sheffield, UK, 10 April 2017.

62. Ojo, A.A.; Dharmadasa, I.M. Analysis of electrodeposited CdTe thin films grown using cadmium chloride precursor for applications in solar cells. *J. Mater. Sci. Mater. Electron.* **2017**, *28*, 14110–14120. [CrossRef]

63. Ng, J.H.G.; Record, P.M.; Shang, X.; Wlodarczyk, K.L.; Hand, D.P.; Schiavone, G.; Abraham, E.; Cummins, G.; Desmulliez, M.P.Y. Optimised co-electrodeposition of Fe–Ga alloys for maximum magnetostriction effect. *Sensors Actuators A Phys.* **2015**, *223*, 91–96. [CrossRef]

64. Allongue, P.; Souteyrand, E. Metal electrodeposition on semiconductors. *J. Electroanal. Chem. Interfacial Electrochem.* **1990**, *286*, 217–237. [CrossRef]

65. Ojo, A.A.; Dharmadasa, I.M. Effect of gallium doping on the characteristic properties of polycrystalline cadmium telluride thin film. *J. Electron. Mater.* **2017**, *46*, 5127–5135. [CrossRef]

66. Akai, T.; Kuraoka, K.; Chen, D.; Yamamoto, Y.; Shirakami, T.; Urabe, K.; Yazawa, T. leaching behavior of sodium from fine particles of soda-lime-silicate glass in acid solution. *J. Am. Ceram. Soc.* **2005**, *88*, 2962–2965. [CrossRef]

67. Dharmadasa, I.M.; Ojo, A.A. Unravelling complex nature of CdS/CdTe based thin film solar cells. *J. Mater. Sci. Mater. Electron.* **2017**, *28*, 16598–16617. [CrossRef]

68. Chaure, N.B.; Young, J.; Samantilleke, A.P.; Dharmadasa, I.M. Electrodeposition of p-i-n type CuInSe$_2$ multilayers for photovoltaic applications. *Sol. Energy Mater. Sol. Cells* **2004**, *81*, 125–133. [CrossRef]

69. Chaure, N.B.; Samantilleke, A.P.; Burton, R.P.; Young, J.; Dharmadasa, I.M. Electrodeposition of p^+, p, i, n and n^+-type copper indium gallium diselenide for development of multilayer thin film solar cells. *Thin Solid Films* **2005**, *472*, 212–216. [CrossRef]

70. Delsol, T.; Samantilleke, A.P.; Chaure, N.B.; Gardiner, P.H.; Simmonds, M.; Dharmadasa, I.M. Experimental study of graded bandgap Cu(InGa)(SeS)$_2$ thin films grown on glass/molybdenum substrates by selenization and sulphidation. *Sol. Energy Mater. Sol. Cells* **2004**, *82*, 587–599. [CrossRef]

71. Ojo, A.A.; Dharmadasa, I.M. The effect of fluorine doping on the characteristic behaviour of CdTe. *J. Electron. Mater.* **2016**, *45*, 5728–5738. [CrossRef]

72. Ojo, A.A.; Dharmadasa, I.M. Electrodeposition of fluorine-doped cadmium telluride for application in photovoltaic device fabrication. *Mater. Res. Innov.* **2015**, *19*, 470–476. [CrossRef]

73. Razykov, T.M.; Ferekides, C.S.; Morel, D.; Stefanakos, E.; Ullal, H.S.; Upadhyaya, H.M. Solar photovoltaic electricity: Current status and future prospects. *Prog. Sol. Energy* **2011**, *85*, 1580–1608. [CrossRef]

74. Kern, W.; Schuegraf, K.K. Deposition Technologies and Applications: Introduction and Overview. In *Handbook of Thin Film Deposition Processes and Techniques*; Seshan, K., Ed.; William Andrew: Norwich, NY, USA, 2001; pp. 11–43.

75. Dharmadasa, I.M.; Samantilleke, A.P.; Young, J.; Boyle, M.H.; Bacewicz, R.; Wolska, A. Electrodeposited *p*-type and *n*-type ZnSe layers for light emitting devices and multi-layer tandem solar cells. *J. Mater. Sci. Mater. Electron.* **1999**, *10*, 441–445. [CrossRef]

76. Ferekides, C.S.; Marinskiy, D.; Viswanathan, V.; Tetali, B.; Palekis, V.; Selvaraj, P.; Morel, D.L. High efficiency CSS CdTe solar cells. *Thin Solid Films* **2000**, *361–362*, 520–526. [CrossRef]

77. Bosio, A.; Romeo, N.; Mazzamuto, S.; Canevari, V. Polycrystalline CdTe thin films for photovoltaic applications. *Prog. Cryst. Growth Charact. Mater.* **2006**, *52*, 247–279. [CrossRef]

78. Cunningham, D.; Rubcich, M.; Skinner, D. Cadmium telluride PV module manufacturing at BP solar. *Prog. Photovoltaics Res. Appl.* **2002**, *10*, 159–168. [CrossRef]

79. Turner, A.K.; Woodcock, J.M.; Özsan, M.E.; Cunningham, D.W.; Johnson, D.R.; Marshall, R.J.; Mason, N.B.; Oktik, S.; Patterson, M.H.; Ransome, S.J. BP solar thin film CdTe photovoltaic technology. *Sol. Energy Mater. Sol. Cells* **1994**, *35*, 263–270. [CrossRef]

80. Ehl, R.G.; Ihde, A.J. Faraday's electrochemical laws and the determination of equivalent weights. *J. Chem. Educ.* **1954**, *31*, 226. [CrossRef]

81. Ojo, A.A.; Dharmadasa, I.M. Optimisation of pH of the CdCl$_2$ + Ga$_2$(SO$_4$)$_3$ activation step of CdS/CdTe based thin-film solar cells. *Sol. Energy* **2018**, *170*, 398–405. [CrossRef]

82. Dharmadasa, I.M. Review of the CdCl$_2$ Treatment used in CdS/CdTe thin film solar cell development and new evidence towards improved understanding. *Coatings* **2014**, *4*, 282–307. [CrossRef]

83. Ojo, A.A.; Dharmadasa, I.M. Optimisation of pH of cadmium chloride post-growth-treatment in processing CDS/CDTE based thin film solar cells. *J. Mater. Sci. Mater. Electron.* **2017**, *28*, 7231–7242. [CrossRef]

84. Hajimammadov, R.; Fathi, N.; Bayramov, A.; Khrypunov, G.; Klochko, N.; Li, T. Effect of "CdCl$_2$ Treatment" on Properties of CdTe-based solar cells prepared by physical vapor deposition and close-spaced sublimation methods. *Jpn. J. Appl. Phys.* **2011**, *50*, 05FH01. [CrossRef]

85. Rios-Flores, A.; Peña, J.L.; Castro-Peña, V.; Ares, O.; Castro-Rodríguez, R.; Bosio, A. A study of vapor CdCl$_2$ treatment by CSS in CdS/CdTe solar cells. *Sol. Energy* **2010**, *84*, 1020–1026. [CrossRef]

86. Schiavone, G.; Murray, J.; Smith, S.; Desmulliez, M.P.Y.; Mount, A.R.; Walton, A.J. A wafer mapping technique for residual stress in surface micromachined films. *J. Micromech.Microeng.* **2016**, *26*, 095013. [CrossRef]

87. Schiavone, G.; Murray, J.; Perry, R.; Mount, A.; Desmulliez, M.; Walton, A. Integration of electrodeposited Ni-Fe in MEMS with low-temperature deposition and Etch processes. *Materials* **2017**, *10*, 323. [CrossRef] [PubMed]

88. Dharmadasa, I.M.; Echendu, O.K. Electrodeposition of electronic materials for applications in macroelectronic- and nanotechnology-based devices. In *Encyclopedia of Applied Electrochemistry*; Kreysa, G., Ota, K.-i., Savinell, R.F., Eds.; Springer: New York, NY, USA, 2014; pp. 680–691.

89. Bhattacharya, R.N.; Oh, M.-K.; Kim, Y. CIGS-based solar cells prepared from electrodeposited precursor films. *Sol. Energy Mater. Sol. Cells* **2012**, *98*, 198–202. [CrossRef]

90. Echendu, O.K.; Fauzi, F.; Weerasinghe, A.R.; Dharmadasa, I.M. High short-circuit current density CdTe solar cells using all-electrodeposited semiconductors. *Thin Solid Films* **2014**, *556*, 529–534. [CrossRef]

91. Ojo, A.A.; Dharmadasa, I.M. Analysis of the electronic properties of all-electroplated ZnS, CdS and CdTe graded bandgap photovoltaic device configuration. *Sol. Energy* **2017**, *158*, 721–727. [CrossRef]

92. Olusola, O.I.; Madugu, M.L.; Dharmadasa, I.M. Investigating the electronic properties of multi-junction ZnS/CdS/CdTe graded bandgap solar cells. *Mater. Chem. Phys.* **2016**, *191*, 145–150. [CrossRef]

93. Sathaye, S.D.; Sinha, A.P.B. Studies on thin films of cadmium sulphide prepared by a chemical deposition method. *Thin Solid Films* **1976**, *37*, 15–23. [CrossRef]
94. Basol, B.M. Processing high efficiency CdTe solar cells. *Int. J. Sol. Energy* **1992**, *12*, 25–35. [CrossRef]
95. Woodcock, J.M.; Turner, A.K.; Ozsan, M.E.; Summers, J.G. Thin film solar cells based on electrodeposited CdTe. In Proceedings of the Conference Record of the Twenty-Second IEEE Photovoltaic Specialists Conference, Las Vegas, NV, USA, 7–11 October 1991; pp. 842–847.
96. Reese, M.O.; Perkins, C.L.; Burst, J.M.; Farrell, S.; Barnes, T.M.; Johnston, S.W.; Kuciauskas, D.; Gessert, T.A.; Metzger, W.K. Intrinsic surface passivation of CdTe. *J. Appl. Phys.* **2015**, *118*, 155305. [CrossRef]
97. Burst, J.M.; Duenow, J.N.; Albin, D.S.; Colegrove, E.; Reese, M.O.; Aguiar, J.A.; Jiang, C.-S.; Patel, M.K.; Al-Jassim, M.M.; Kuciauskas, D.; et al. CdTe solar cells with open-circuit voltage breaking the 1 V barrier. *Nat. Energy* **2016**, *1*, 16015. [CrossRef]

MDPI
St. Alban-Anlage 66
4052 Basel
Switzerland
Tel. +41 61 683 77 34
Fax +41 61 302 89 18
www.mdpi.com

Coatings Editorial Office
E-mail: coatings@mdpi.com
www.mdpi.com/journal/coatings

9 783039 430406